The Engineering Design Challenge:

A Creative Process

Synthesis Lectures on Engineering

Essentials of Applied Mathematics for Scientists and Engineers
Robert G. Watts
2007

Project Management for Engineering Design
Charles Lessard and Joseph Lessard
2007

Relativistic Flight Mechanics and Space Travel
Richard F. Tinder
2006

The Engineering Design Challenge: A Creative Process
Charles W. Dolan

ISBN: 978-3-031-79356-1   paperback
ISBN: 978-3-031-79357-8   ebook

DOI 10.1007/978-3-031-79357-8

A Publication in the Springer series
SYNTHESIS LECTURES ON ENGINEERING

Lecture #21
Series ISSN
Synthesis Lectures on Engineering
Print 1939-5221   Electronic 1939-523X

The Engineering Design Challenge:

A Creative Process

Charles W. Dolan
University of Wyoming

SYNTHESIS LECTURES ON ENGINEERING #21

ABSTRACT

The *Engineering Design Challenge* addresses teaching engineering design and presents design projects for first-year students and interdisciplinary design ventures. A short philosophy and background of engineering design is discussed. The organization of the University of Wyoming first-year *Introduction to Engineering* program is presented with an emphasis on the first-year design challenges. These challenges are presented in a format readily incorporated in other first-year programs. The interdisciplinary design courses address the institutional constraints and present organizational approaches that resolve these issues. Student results are summarized and briefly assessed. A series of short intellectual problems are included to initiate discussion and understanding of design issues. Sample syllabi, research paper requirements, and oral presentation evaluation sheets are included.

KEYWORDS

engineering, design, challenges, first-year, interdisciplinary, multidisciplinary, assessment, outcomes, organization, evaluation

To M for years of understanding, help, and support

Contents

Foreword

If you were to teach engineering design to first-year undergraduate students, how should you proceed? This question is at the heart of the present book by Professor Charles Dolan, an accomplished designer and adept educator.

The question may seem daunting, as engineering design entails the interplay of creativity and specialist expertise gained from a rigorous path of engineering study. First-year undergraduate students may be naturally creative, but they are just starting down the path and thus unlikely to have the know-how needed for meaningful engineering design. Yet as Professor Dolan shows, the design process essentially entails a series of basic considerations that first-year students can readily grasp.

Professor Dolan's design challenge requires that students organize themselves into teams to pursue a given design objective subject to a set of prescribed design constraints (rules, as he calls them). To attain the design objective, each team must prepare a simple plan, coordinate and record its activities, performance test the resulting design, and write a concise technical report. The challenge provides students with a substantial foretaste of the design process, and enables them to better anticipate the knowledge needed for successfully undertaking more complex engineering designs.

Not satisfied with tackling one major question about teaching design, Professor Dolan then shifts gears to take on another: How to teach interdisciplinary design to a team of undergraduate students drawn from different engineering disciplines?

Addressing this question can be complicated by non-trivial curricular and administrative issues, as he explains. Additionally, it can be complicated by the issue of conceiving design projects that meaningfully involve students from different engineering disciplines over a single semester. Nevertheless, Professor Dolan successfully addresses all of these issues, to the extent that one of his interdisciplinary projects, *Disappearing Roads*, received national recognition and was implemented at sites requiring a minimal construction footprint.

Professor Dolan's design expertise draws on his vast experience as a consulting engineer working with teams to design a broad range of concrete structures, including a remarkable series of concrete monorail structures where design requirements were subject to comprehensive sets of constraints pertaining to sound structural performance, transportation efficiency, convenience, economics, and aesthetics. He has coauthored a leading book on the design of concrete buildings, engaged in numerous research projects, and for many years has been a member of the country's (and arguably the world's) principal committee prescribing the design rules for concrete buildings.

He teaches in an engaging and anecdote-rich manner, and occasionally enjoys pointing out less common factors that sometimes enliven the design process. For example, I recall him recounting with some humor the challenge he once faced as a member of an international design team whose members spoke four languages including Norwegian and Japanese. Great care was needed to ensure clarity of communication, a point he often emphasizes to students.

Professor Dolan is a "designer par excellence." The University of Wyoming has been fortunate to have him as its founding H. T. Person Chair of Engineering Education, which was established for the express purpose of attracting world-class engineers and engineering educators to help teach the design process to undergraduate engineering students. Professor H. T. Person himself was an accomplished teacher extensively recognized for his contributions to engineering practice in Wyoming and surrounding regions.

This book is Professor Dolan's leave-taking contribution to engineering education. It offers a useful, brief digest of his experience teaching design and introducing engineering to first-year students. In late 2012 he retired from the H. T. Person Chair, and now works at his own pace on a variety of very interesting engineering design and research activities.

Robert Ettema
Dean and Professor
College of Engineering and Applied Science
Laramie, Wyoming
March 2013

Preface

Hjalmar Thorval (H. T.) Person was Professor, Department Head, Dean, and President of the University of Wyoming in the period from 1929 to 1968. Known as "Prof," he was an accomplished teacher, sometimes teaching 20 credit hours per semester. As dean he was instrumental in moving the College of Engineering and Applied Science from its infancy to a nationally recognized program. As president, he is credited with quelling faculty dissention on campus and returning the university to a sense of progress. A Fellow of the American Society of Civil Engineers, he served as a director, on the executive committee, and on technical committees including Drainage and Irrigation, and Registration of Engineers.

H. T. worked summers as a practicing engineer for American Bridge Company and the Missouri Highway Department in order to bring state-of-the-art design to his classes. In Wyoming he served variously as the director for the U.S. Coast and Geodetic Survey and the State Control Survey. Person was the state's chief negotiator on the Upper Colorado, Yellowstone, Cheyenne, Snake, Niobrara, and Columbia River Compact Commissions. He was appointed to the President's Missouri River Basin Survey Commission and in 1965 was named to the Upper Colorado River Commission. In a region of the country where "whiskey is for drinking and water is for fighting" the commitments to and importance of these commissions were both time consuming and essential to the State of Wyoming. He was recognized for his service by the National Council of State Board of Engineering Examiners, and the Four-State Irrigation Council. H. T. Person received the first Wyoming Society of Professional Engineers' "Engineer of the Year" award and the University of Wyoming's "Medallion of Service."

In the early 1990s, several alumni joined to create an endowment to establish a chair to honor the vision of H. T. Person and the "Prof's" dedication to undergraduate education. Over 200 individuals, groups, and foundations supported the endowment. I had the distinct pleasure to discuss H. T. Person's legacy with several of the leaders of this effort including Gus Albert, "Tut" Ellis, Harold Kester, Albert "Boots" Nelson, Frosty Kepler, and Ken Kennedy. To a person, they all expressed admiration for the contribution that H. T. made to their careers and the support he offered to the multitude of students studying at the university.

As the endowment was being initiated a series of lectures in H. T. Person's honor was established. Each fall the college invites a noted individual to be the H. T. Person Distinguished Lecturer to address our alumni, students, and faculty as part of the university's homecoming weekend. The selection of the speakers is an opportunity to present timely topics and, in many cases, to highlight the accomplishments of our alumni. For example, when the movie *Apollo 13* was released, Mr. David Reid, a graduate of the University of Wyoming Mechanical Engineering program and flight controller for the Apollo 10, 11, 12, 13, and 14 missions, spoke on the real issues of bringing Apollo 13 back to earth. Mr. Larry Novak of the design firm Skidmore, Owens and Merrill, spoke on the rescue efforts at the World Trade Center in 2002. In 2010 Mr. Joe Leimkuhler, Manager of Offshore Well Delivery for Shell Oil, spoke on oil drilling operations following the Deepwater Horizon oil spill. Complementing these lectures, speakers meet with classes to discuss the projects in detail and to provide insight on the value of engineering education. Even though H. T. Person was a civil engineer, a conscious effort is made to select speakers from all engineering disciplines. Thus, students and faculty have an opportunity to explore new ideas and concepts. A complete list of the H. T. Person distinguished speakers is found in Appendix I.

In 2000, Mr. John Clark, noted bridge engineer, spent a semester on campus as the H. T. Person professor in residence. He supplemented the H. T. Person Homecoming Lecture with a presentation on the collapse of the Quebec River Bridge. Mr. Clark brought the background of the bridge design and construction and the Order of the Engineers ceremony together for the students. He challenged and exposed the students to the high level of responsibility that is expected of them as practicing engineers. The successful completion of Mr. Clark's professorship suggested that a permanent chair would best suit the vision for improved undergraduate education.

In 2002 a national search was conducted for the first permanent H. T. Person Chair. I received the appointment and have had the privilege to focus on undergraduate education and engineering design. It has been a true pleasure to be able to share this endeavor for others to use. My years of professional practice have been essential to my ability to provide meaningful and relevant experiences.

The challenges presented here are the culmination of over a decade of development. When they began, the college had an embryonic first-year engineering experience focusing on the student transition from high school to the university. Therefore, initial efforts of the H. T. Person Chair focused on the first-year Engineering Design Challenge. After three years, the *Introduction to Engineering* course was well established and was being managed by Dr. Thomas V. Edgar. The design and preparation of the annual first-year Design Challenge remained in the purview of the H. T. Person Chair. Dr. Edgar prepared many of the common lectures for the course and his collaboration on this course provided me opportunity to initiate interdisciplinary design courses. His keen insight and ability to involve students was a particular asset.

The Interdisciplinary Senior Design program courses were created to be truly broad in scope. The completed projects are discussed in detail as is the philosophy and organization to make them work. In 2010, Dean Robert Ettema and I discussed the possibility of creating this volume to transfer the experience of developing undergraduate design programs to others who are interested in similar ventures. While this effort focuses on the work of the undergraduate students at the University of Wyoming, the concepts are transferrable. As is the case in most educational situations, motivated students and faculty rise to the challenge. More than anything, this book is a testament to the ability of the students to accept the challenges.

As John Donne's poem reads, "No man is an island entire of itself …" such is also true of this volume. As with any venture, there are many colleagues contributing to these projects. Dean O. A. Plumb and Dean Robert Ettema have supported the sometimes non-conventional approaches to developing design efforts. Associate Deans for undergraduate education, David Whitman, Richard J. Schmidt, and Steven F. Barrett were always available for consultation on methods to assess progress and to issue the semester-end student evaluation surveys. Dr. David Mukai and Dr. Jennifer Tanner have been valuable colleagues and co-PIs for research endeavors.

Most important is the concept of a chair devoted to undergraduate teaching. The financial and moral support of the original donors to the H. T. Person Endowment is deeply appreciated. Four people in particular have served on the H. T. Person Advisory Board and have been sounding boards for some of the ideas developed in this program. Albert "Boots" Nelson has been a constant source of inspiration and support of these efforts. Tom Lockhart, Floyd Bishop and, Bill Bellamy have also provided ideas and insight for these efforts. In addition to their support of undergraduate education, it is a tribute to their professional careers and foresight that several of the original donors are members of the College of Engineering and Applied Science Hall of Fame. I thank them all for their active interest in engineering education even as they pursued other interests in their careers and retirement.

Charles W. Dolan
H.T. Person Chair of Engineering
Laramie, Wyoming
March 2013

CHAPTER 1

The World of Engineering Design

When Jacob Boorstin, formerly head of the Library of Congress, released his book *The Discoverers*,[1] there was no mention of any engineering feat. *The Discoverers* examines the intense and often individual pursuits of people driven to understand the world around them. Beginning with the ancient Greeks and progressing through Newton to Watson and Crick, *The Discovers* presents a personal quest to understand science and the natural world. As such, science requires the intellectual capacity to examine a multitude of data and assimilate that information into a coherent theory. When successful, each theory can be replicated and validated by others.

It wasn't until Boorstin's publication of *The Creators*[2] that engineering was acknowledged. In *The Creators*, art, music and sculpture appear alongside works in stone, concrete, and steel as testament to human creativity. Michelangelo, Monet, and Mozart share space with DaVinci, Brunelleschi, and Eads. While engineers do not typically see themselves in the same venue as the great artists, composers, and playwrights, they share a common trait. They create something where nothing existed.

Engineering design is a unique activity with a single problem statement and multiple solutions. Often only a small number of the solutions come close to meeting all of the project requirements. In *Path Between the Seas*,[3] David McCullough describes the extraordinary difficulties of working in the mud and landslides during excavation of the Panama Canal. When John Stevens, J. J. Hill's chief engineer of the Great Northern Railroad, was put in charge of the construction, he immediately saw a railroad problem not an excavation problem. By employing the steel rails instead of working in the constantly changing mud, Stevens was able to stabilize the construction and move the project toward a successful conclusion. *Path Between the Seas* additionally places engineering design into the larger context of societal needs. Less than 10 percent of the book deals exclusively with the technical problems while the remainder deals with the politics, organization, and personalities engaged in this world class project.

1.1 DESIGN RECOGNITION

If engineering design is such a creative process, why are so few engineers recognized for their endeavors? This question evokes two different and somewhat diametric responses. The first argument is that engineering practice has not engendered the cult of personality evident in the arts, architecture, and scientific communities. The second argument is far more germane to the development of the craft.

Engineers work in teams with more emphasis on the end result than on the individual contribution. Simply put, individuals are easier to recognize than teams.

The word Science derives from the Latin *scientia*, meaning "knowledge." Using the Wikipedia description, science "is a systematic enterprise that builds and organizes knowledge in the form of testable explanations and predictions about the universe." Scientific theories and discoveries, unlike design projects, are often named for the discoverer not the principle involved. Newton's laws of motion and Boyle's law for gas pressure are examples of named scientific principles. An examination of the development of the atomic bomb at Los Alamos, New Mexico, illustrates the counter practice. Everything from the uranium separation to the development of the explosive detonation devices are engineering endeavors. That is, scientific principles are used to develop practical applications to meet project criteria. Yet, by definition, there were no engineers at Los Alamos, only scientists and "research scientists."[5]

While Henry Petrosky, in *Remaking the World: Adventures in Engineering*,[4] decries the lack of engineering recognition, the profession identifies the larger issue as the need to be able to communicate with each other and with the public at large. Engineering education places a premium on teamwork and oral and written communication. Articulation of complex issues, solutions, and the impact on society fall to the engineer. History has proven there is no place for hubris in the engineering design profession. One needs only to read the impact on Robert Mulholland after the collapse of the St. Francis Dam in California, to fully appreciate the impact on the individual and society resulting from neglect or oversight on design.[6]

1.2   WHAT IS ENGINEERING DESIGN

Design is a creative process. As such, the definition of design changes with its associate profession. Art, engineering, music and all creative enterprises have design components. For example, civil engineering design may include creation of contract documents, calculations, or studies that describe how to assemble materials in new or novel ways. Similarly, computer designers may create a new generation core chip just as a musician drafts a score or an artist sketches the start of a painting. An example of the creative symbiosis of technology, design, and the arts is Wikipedia, which is why it is used in this context. The platform allows all to assemble information and at the same time contains a series of checks and balances to assure validity of the work.

More formally, Wikipedia defines design as follows:

(noun) a specification of an object, manifested by an agent, intended to accomplish goals, in a particular environment, using a set of primitive components, satisfying a set of requirements, subject to constraints;

(verb, transitive) to create a design, in an environment (where the designer operates)

This formal definition provides a framework for the development of engineering design challenges presented in this book. Specifically, the design must satisfy a set of criteria or requirements and is constrained by external factors such as available materials or cost.

1.3   WHY TEACHING DESIGN IS DIFFICULT

The idea that design creates something where nothing previously existed is a source of anxiety when teaching design. In the university environment, courses are offered by individual topic and subject area, which are often aligned with the professor's personal and professional interest. In most cases, the subject material is further structured to be narrow and analytic in nature. "Analytic" implies the student assesses the information provided, prepares the necessary calculations, and presents a unique solution. This is especially true in the developmental years of engineering education. In these courses, the student is assembling the building blocks needed for the rest of his or her career. At the same time, coordination between these core courses is minimal or non-existent. In consequence, students have several years of experience and practice generating a single solution to a well constrained problem, which is then graded by the faculty member as right or wrong. Throughout this process, thinking that integrates coursework from multiple subjects is generally absent or not developed.

Somewhere late in the junior year, or certainly in the senior year, the engineering or computer science student runs into the "design project." Depending on the resources available, faculty members often simplify the design practice questions. Simplification is for self-survival. In a class of 30 to 100 students there will be a wide array of acceptable solutions. Grading this multitude of solutions engages a substantial amount of the faculty member's time and is often not suitable for delegation to a "teaching assistant." Even if a teaching assistant is available to aid with the grading, the faculty member is responsible for defining the range of acceptable solutions and needs to provide insight on the "correctness" of the solution domain.

Developing engineering design problems that require substantial interaction with a student group is even more challenging. These challenges range from introductory first-year courses to interdisciplinary senior design courses. In each case the logic, organization, and assessment of the design courses are presented.

1.4   THE ENGINEERING DESIGN CHALLENGE

The "Engineering Design Challenge" is then a paradox. To the educator, the challenge is to impart concepts of design to students steeped in a tradition of solving analysis problems. To the student design is an uncomfortable step from known to unknown conditions, often with apparently sketchy definitions of the problem requirements. The following is an examination of both sides of the paradox. Each section discusses the design challenge presented to the students and the objectives and trials of managing the design challenge effort.

Complementing the Engineering Design Challenge is an overview of the University of Wyoming first-year *Introduction to Engineering* course. This course is required of all first-year engineering students and includes the first-year design challenge. The success, failure, and lessons learned from this course are discussed.

REFERENCES

[1] Boorstin, D. J., *The Discovers*, First Vintage Books, New York 1983.

[2] Boorstin, D. J., *The Creators: A History of Heroes of the Imagination*, First Vintage Books, New York 1993.

[3] McCullough, David G., *Path Between the Seas: The Creation of the Panama Canal, 1870–1914*, New York, Simon and Schuster, 1977.

[4] Rhodes, Richard, *The Making of the Atomic Bomb*, Simon & Schuster, New York, 1986.

[5] Petroski, Henry, *Remaking the World: Adventures in Engineering*, First Vintage Books Edition, New York 1997

[6] Reisner, Marc. *Cadillac Desert: The American West and its Disappearing Water*, New York, Penguin Books, 1993.

CHAPTER 2

Introduction to Engineering

2.1 ES 1000 INTRODUCTION TO ENGINEERING

Each fall semester, the University of Wyoming College of Engineering and Applied Science offers 13 sections of *ES 1000 - Introduction to Engineering*. With 22 to 30 students per section, this is the largest common course in the university. Instructors are typically tenured faculty or extended term lecturers who volunteer to lead the course. The professor is assisted by an undergraduate engineering peer assistant. A common course syllabus is developed and a set of prepared lecture notes accompanies the syllabus to assist each professor. A sample course syllabus is included in Appendix II. Each fall a design challenge is developed and presented to the first-year students.

The University of Wyoming *Introduction to Engineering* course is one credit hour and is intended to cover a variety of topics that are required by the University Studies Program (USP) as well as introducing the students to a design exercise. Embedded within the ES 1000 course are the university requirements for information literacy and intellectual community. Both of these requirements are somewhat loosely defined. For example, the University Studies Program defines an intellectual community:

Intellectual Community Definition:

Courses that fulfill the Intellectual Community requirement of University Studies provide students with an introduction to the purpose and philosophy of higher education. These academic, content-based courses, designed for first-year students, focus on the critical-thinking skills necessary to understand, analyze, and produce knowledge within the framework of the discipline or area of inquiry in which the course is offered.

In attempting to address all areas of study within the university, this definition is more of a description than a definition and offers only marginal guidance for the organization of the course. Similarly, the University Studies Program defined information literacy from the American Library Association.

Information Literacy Definition:

Information Literacy, as defined by the American Library Association, is the ability to "recognize when information is needed and to locate, evaluate and use effectively the needed information."

These two requirements are complementary to the design process. Even so, combining all of these actions into a single course, let alone a single credit hour, is challenging. Considering the above, the university catalog description of the ES 1000 course is as follows:

> *ES 1000. Orientation to Engineering Study. 1. Skills and professional development related to engineering. Involves problem solving, critical thinking and ethics, as well as activities to help transition to university environment. Required of all freshmen entering engineering curricula. Students with credit in UNST (University Studies) 1000 may not receive credit for this course. (Normally offered both semesters)*

The *I* and *L* designation in the catalog description indicate that the course meets the intellectual community and information literacy requirements and are approved for such by the USP Committee. With only a single credit hour to accomplish the objectives of this myriad set of requirements, the course content is relatively packed. Over the years, the course has evolved from an effort to assist engineering students transitioning from high school to college life to a course with substantial engineering and design content. The current content primarily addresses the philosophy of engineering while engaging the student in the engineering profession. Approximately 10 years ago the course was modified to meet both the 2003 USP criteria listed above; it also added a design challenge to the curriculum. In addition to the course professor, each section is assigned a peer assistant, who receives a small stipend. The peer assistant assists the faculty member in the presentation of the course and becomes a contact for the first-year students. The peer assistant concept has proven to be exceptionally valuable. The peer assistant is able to answer a number of transitional questions that faculty members are, by and large, unaware of or unable to address. These questions deal with issues such as campus food, dating, roommate problems, selection of faculty members for courses during advising, and related topics.

When broken down to its fundamentals, ES 1000 has three important components. The first component introduces the students to campus life. The second component requires students to become active in collaborative work, primarily by establishing groups engaged in the design challenge. The third component, information literacy, is an individual effort that requires the students to prepare a paper on a topic related to the design challenge and a second paper assessing the sources used in the research paper.

2.1.1   INTRODUCTION TO UNIVERSITY LIFE

The introduction to university life component engages students in the College of Engineering and Applied Science and the campus at large. There are a required number of activities that each student must attend. These include professional society meetings within the college and events external to the college. Each student is asked to attend the senior design symposium presentations at the end of the semester. Participation is recorded by a self-reporting system structured to foster responsibility and communication.

A critical piece of this component is encouraging students to participate in professional society activities. An underlying philosophy for this requirement is to retain students. Retention is improved if the student is engaged in their areas of interest. Undeclared students visit professional societies in areas of interest. In addition to the regular professional societies, the students may also select from the Society of Women Engineers, the Minority Engineering Program, and Engineers without

Borders. Through this effort students meet upper-class students and become comfortable both in the college and their area of study.

2.1.2   INTELLECTUAL COMMUNITY

The intellectual community requirement engages the students through the design challenge. While the students work in small groups, the challenge is organized to reward collective efforts. In the detailed challenges that follow, the entire section is taken as a team. This structure supports the exchange of ideas and concepts within the class rather than developing an attitude of secrecy and exclusion.

To encourage cooperative endeavors, students are required to present oral summaries on the progress of their design efforts. Typically, two presentations occur during the semester. The first presentation is during the preliminary design phase. This presentation opens the array of potential solutions to the design challenge or focuses on the development of one aspect of the design. The second presentation either follows the research paper or the design challenge. If the presentation is on the design challenge, then the students must explain why their designs worked and what aspects of the design could be improved. If the oral presentation is on the research paper, the students correlate their research to the solution they developed for the design challenge. The selection of the design challenge or the research topic for the challenge is at the discretion of the section professor. The selection of an option in any given semester depends upon the scheduling of the design challenge. Because the design challenge often requires the use of particular buildings or facilities on campus, the presentation date often has to be coordinated with other academic units. When the challenge occurs at the very end of the term, the research topic is selected for oral presentation.

To engender building an intellectual community, each group is required to develop a design notebook. As motivation for developing and maintaining a notebook, the work of College Hall of Fame member Thomas Osborne is used as a case study. Osborne worked for Hewlett Packard and was involved in the design of handheld calculators. Osborne's notebooks at HP were instrumental in HP winning a lawsuit and receiving the patent for the first fully functional engineering calculator. A thought-provoking interview with Osborne can be found on the web.[1,2,3]

The Design Notebook

To assess student engagement, each group is required to maintain a design notebook. Bound notebooks, e.g., spiral or similar volumes, are preferred, however, loose leaf notebooks are allowed to accommodate the students' conflicting schedules. The notebook is started early in the semester to record ideas, thoughts, and designs. The notebook must contain the following information:

- A title page containing the design group name and the name and email of each group member.

- A summary page, following the title page, with the following critical information:

    - Page numbers for the testing and evaluation program.

    - A brief summary of the test data results.

  – The total expenditure to construct the project including a statement that the budget restrictions were met.

- Pages must be numbered and dated.

- Pages should include components of the design work including: sketches, references, notes, calculations, list of materials, costs, alternate materials considered, summaries of group discussions, conclusions and decisions, and any other activities relevant to the design challenge. Comments on ideas that work, things that didn't work, and changes to initial design concepts are appropriate.

- The design notebook is turned in at the design challenge.

- The design notebook is source material for the oral presentation.

- The final entries should include comments on how to improve the design.

Notebooks are reviewed by the peer assistants at the registration for the design challenge. They are graded on a three-part scale. Zero points are assigned for no notebook, 5 points for a fair to poor compilation, 7 points for an average submittal, and 10 points for a thorough effort. All members of the group receive the same grade.

2.1.3 INFORMATION LITERACY

The information literacy component of ES 1000 requires the students to prepare a brief paper and to acquire the skills to critically assess information sources. Students examine a number of different information sources and evaluate the quality of those sources. The paper requires the students to evaluate relevant sources and limits the number of references allowed. The limitation of references forces students to concentrate on those references most valuable to presenting their topic. Each student is asked to locate one reference from a peer-reviewed journal, one reference from popular literature, one reference from the Internet, and one reference in opposition to the position they are presenting. A critical part of the information literacy paper is explaining to the students how to differentiate between Internet sources and peer-reviewed articles since both appear on the Internet. With the availability of search engines in the University Library, and the accessibility of Google Scholar, the students must be able to differentiate between general Internet-based sources and peer reviewed material.

A second paper asks the students to assess their sources. Restricted to two–three pages in length, the exercise forces students to critically examine the material they were presenting and create a rationale for the validity of their source material. The students submit a portfolio of the articles that they used for their paper. The portfolio accomplishes two objectives. The first objective demonstrates to the student that resource material should be archived for their own use. The second objective allows the professor to verify that the source requirements are satisfied. A full description of the information literacy requirements and the assessment paper requirements is provided in the Appendix III.

Perhaps one of the more interesting outcomes of this exercise occurs when students comment on how difficult the journal papers are to read. Comments like this provide an opportunity to reinforce the value of the education, why the curriculum is structured to build the fundamentals of and engineering education, and promote an awareness of the value of professional papers compared to news sources.

2.2   ROLE OF H. T. PERSON CHAIR

The H. T. Person Chair of Engineering position was made possible by University of Wyoming Alumni who recognized the value of a strong emphasis on engineering fundamentals and desired to make a serious contribution to undergraduate education. An annual task of the H. T. Person Chair is to establish the first-year design challenge. The endowment also provides a small amount of discretionary funds to support these challenges.

The H. T. Person Chair job description is 60 percent teaching, 5 percent advising, 5 percent service, and 30 percent research and creative endeavors. Historically, two–three credit courses and one section of ES 1000 are taught in the fall and 2-three credit courses are taught in the spring. Honors courses are often taught as a voluntary overload. Two to four graduate students are directed each year.

In retrospect, managing the interdisciplinary senior design courses require considerable effort. In addition to the course organization, these courses require coordination with faculty in related fields so expertise is available when needed. A reasonable work load would be to teach only the interdisciplinary course in a given semester.

2.3   CLOSING COMMENTS ON ES 1000

2.3.1   STUDENT ENGAGEMENT

Prior to 2003 the student response to ES 1000 had been relatively lackluster. To reinforce the importance of the course and to emphasize its place in the University Studies Program, in 2003 the college instituted a Freshman Convocation. The Convocation is held on the Monday of the first day of class with ES 1000 classes beginning on Tuesday. At the Convocation, the Dean welcomes the students and presents the importance of the class, the faculty members and peer assistants are introduced, and the design challenge for the semester is unveiled. The convocation improved student engagement significantly.

2.3.2   THE COURSE STRUCTURE

After the first cycle of teaching ES 1000, it was apparent that a typical one-hour schedule had to be restructured to meet the demands of the program. The course dragged on and students lost interest toward the end of the semester. Consequently, the course was modified to meet twice a week for half a semester. The modified schedule enhanced student engagement and eliminated conflicts between the design challenge and final term projects in other classes.

2.4 ASSESSMENT

ES 1000 is successfully engaging students in the college. The course provides the students with critical tools needed to advance their education and a myriad of topics and activities to explore. The expanse and complexity of an engineering education is presented and enables students to assess early in their career if this course of study is appropriate for them. The grading structure of ES 1000 balances class attendance, the design challenge, and outside activities. Adoption of a similar program can adjust where the emphasis is placed in the grading system.

REFERENCES

[1] www.hp9825.com/html/osborne_s_story.html

[2] www.viddler.com/explore/sleibson/videos/4/

[3] www.edn.com/electronics-blogs/4306814/how-hp-got-its-first-calculators-video-interview-with-tom-osborne

CHAPTER 3

First-year Design Challenge Development

3.1 OVERALL CHALLENGE PHILOSOPHY

An overriding concern in engineering educations is engineering and applied science students do not see design until their junior or senior years. This is a long time to wait and an opportunity for students to lose interest. The first-year design experience is to motivate students to realize engineering can be fun as well as complex. The details of first-year engineering challenges are provided in the following chapter. The philosophy and underlying assumptions of the challenges are presented here.

A review of some design challenges began to show the difficulty in designing challenges for large groups of students. The MIT "King of the Hill" challenge and the University of Oklahoma "Pumpkin Toss" demonstrate excitement and student engagement. At the same time, they are a competition, complex for first-year students, and involve a substantial financial outlay. While life itself is a competition, engineering generally requires collaboration. Therefore, challenges, not competitions, are developed to promote a cooperative endeavor. The challenges are sensitive to the fact that college and students' budgets are severely constrained. This leads to a second goal of these design challenges: limit the budget for the project.

In several of the following challenges, parts are identified and supplied to the students. Funding for these parts comes from the H. T. Person endowment. These parts were supplied to balance the playing field by requiring all groups to work with a common set of components. The motors are typically underpowered or have a high a RPM in their native mode. Thus, each group has to adjust their design to the materials available.

One global objective of the challenge is to minimize the rules. This makes the challenge a true outcome-based activity where creativity and innovation are rewarded. For each challenge, questions arise that require clarification. These questions are classified as "proprietary" or "public" by the faculty member in charge or by the student request. Public questions request general clarification of the rules and are answered on a FAQ site set up on the ES 1000 class website. For example, a public question might be: "Can I change the batteries after each run?" Proprietary questions deal with groups wanting to know if an aspect of the design is allowed. An example of a proprietary question that is used in class is the keel design of the Australian America's Cup yacht several years ago. The winged keel design was approved but kept secret until the race and the Australian yacht ended up with a significant competitive advantage. Students identify their request as public or proprietary and, if private, they are responded to individually. An example of a proprietary question might be: "Can we use CO_2 cartridges for power?" In the case of the CO_2 cartridge, the response may also ask for a safety plan on the cartridge use.

The challenge envisions designs that can be completed in the student dorm room. The college machine shop has space available for the first-year students. The area contains a number of hand tools and drill presses. Students must attend a shop safety video before gaining permission to use shop facilities. Shop times are scheduled when shop staff supervision is available.

3.2   SAFETY

The first challenge assumed that the students would behave in a safe, responsible manner, especially following the lectures on ethics and holding public safety paramount in professional engineering efforts. On the day of the challenge, one group launched a vehicle powered by 16 bottle rockets on the ballroom floor of the Student Union! Therefore, the following safety clause became part of every challenge.

> *The objective of the challenge is to foster engineering creativity and cooperation. The design group is responsible for ensuring safety of participants and spectators during the challenge. Groups using any feature deemed dangerous by the judges may be asked at any time to prepare a safety plan or suitably modify the design before continuing in the challenge. Offending designs may be disqualified at the discretion of the faculty member or peer assistants. Use of pyrotechnic or similar devices is strictly prohibited. Any questions regarding safety may be directed to your professor.*

The safety statement is general and intended to remind the students of their obligations. Occasionally, designs come forward that are on the margin of safe operation. In these cases, the group is required to prepare a safety program that must be approved before their design is allowed to participate. Requiring a safety plan has two outcomes. Either the design is modified so a plan is not needed or the students prepare a plan and learn another important aspect of design development.

3.3   COST AND TIME CONSTRAINTS

Engineering design is driven by cost effectiveness. To maintain this philosophy in the challenge, students' budgets were initially set at \$10–\$15 per group. "Free" materials could be used; however, the definition of "free" needed explanation.

> *The definition of "free" is that it has no commercial value. In short, the professor can elect to keep it or throw it in the trash following the competition. "Rented" or "borrowed" materials are not allowed.*

The "free" aspect of the challenge is to encourage the students to use materials that may be commonly available or otherwise scrap. Plastic drink containers, cardboard, and similar materials are readily available and have successfully incorporated into many student designs.

The "borrowed or rented" clause is added because students are very skilled in "gaming the system." In one design challenge, a group showed up with a $20,000 piece of robotic equipment "borrowed" from a parent's company. While the design was certainly creative, it was not in the spirit of economical design or the design challenge.

The design challenge has students working in groups of two or three. Experience with these challenges suggests that providing kits is not needed when parts are readily available locally. Challenge budgets increased to $30–$50 per group when no parts were provided. This is incorporated into the class structure as appropriate.

The class schedule also constrains the budget. With only seven to eight weeks from start to finish, there is not sufficient time for students to execute a full set of design plans, prototypes, and finished products. The quality of the student designs improves when trial runs are required. Time to repair or upgrade the designs is built into the schedule. The ideal schedule has one to two weeks between trial testing and the final challenge. In cases where trial runs can be completed and initial data recorded in a continuous manner, less time is needed between the trial and the challenge.

An important parameter in determining the schedule for trial runs and the challenge is the possibility of damage to the design from the challenge testing. Thus, the car side impact and the Styrofoam airplane flight require a longer period between testing and the challenge as the design may be damaged in testing. Awareness of these limitations allows the schedule to provide ample time for modification or replacement.

3.4 AWARDS AND PRIZES

To minimize the competition aspect of the challenge and to reinforce collaboration, individual prizes are not generally offered. Pizza parties for the section with the best overall performance are often provided. The latter is consistent with encouraging cooperation among the groups in each section.

To assess whether prizes were an effective inducement to improve the design, on two occasions, the author provided cash prizes for the best design performance. In one case, the carpet climb, this was extremely effective inducement. In another case, the underwater recovery vehicle, it led to a chaotic situation with less than professional behavior. Keeping awards at the section level was most effective.

3.5 ASSESSMENT

The design challenge is very popular and is often cited as the most interesting and memorable part of the first-year experience. At the end of each semester, the professors and peer assistants gather and critique the challenge, and a semester-end survey is issued to each student. These closing comments are based on student reviews and faculty and peer assistant debriefings.

- The low-cost, open-ended design challenges are very popular.

- The design challenges are an effective introduction to the engineering program.

- On a typical challenge, approximately 20 to 30 percent of the groups will successfully complete the challenge. The toughest challenge, the carpet climb, had 2 of 134 groups succeeded.

- Requiring trial runs as part of the challenge improves the overall design by effectively requiring the groups to complete their design prior to the actual challenge.

- The budgets listed on each challenge are typically sufficient.

- Access to the shop or to basic hand tools is a benefit but not absolutely necessary.

- Challenges can be used by individual sections or by an entire class.

The one minor deficiency in the ES 1000 program is the lack of time spent working with the students on formal design development. There are two reasons for this. First, the total number of topics to be covered diminishes the time available for design. Second, with 13 sections and typically 10 different professors, a lack of consistency on the design effort between sections is possible. A pre-semester briefing helps assure all faculty and peer assistants understand the objectives of the challenge. Faculty members have considerable freedom to then adjust the course, but not the challenge rules, to suit their own goals. The syllabus in Appendix II illustrates the overall course and design-directed activities.

CHAPTER 4

The First-year Design Challenges

4.1 INTRODUCTION

This chapter presents the design challenges developed at the University of Wyoming. The challenges are updated to describe the facilities needed for the design challenge, to incorporate relevant "Frequently Asked Questions," and to incorporate results from running the challenges.

The issue of facilities is not trivial. The University of Wyoming is fortunate to have widespread cooperation among various facilities within the university community. At the same time, gaining access to the indoor football practice field requires coordination between the College of Engineering and Applied Science and the Athletic Department. One consequence of this cooperation is that the design challenge is typically held in October during an away football game. This timing provides the challenge with wider access to campus facilities and less competition for attention but also means that the syllabus is modified each year to synchronize with the facilities needed for the challenge.

Consideration of the facilities is additionally tied to visibility of the engineering program. When possible, the challenges are held in highly public areas. The atrium of the campus library was an ideal setting for the carpet climb. The Student Union ballroom served for the slalom challenge. Having an audience augments the experience. If the challenge is held in a public area, a poster or other description of the challenge assists the public in understanding the activity.

Each challenge describes the challenge objective, the rules for the challenge, the safety statement, the challenge organization, and concluding comments. The challenge and rules statements vary depending on the constraints placed on the challenge and available facilities. The organizational discussion addresses preparation, evaluation, and peer assistant activities required prior to the semester, during the semester, and on the challenge day.

If the challenge is subsidized, the subsidy is discussed in the challenge description. Subsidies often take the format of providing common components that must be used in the challenge and are funded by the H. T. Person Endowment. The source of these components is identified, although some caution is offered because the individual parts are not uniformly available from year to year. Subsidies are typically less than $1,000 spread over the 13 sections of first-year students.

The challenge is typically completed on Saturday morning, although one challenge was successful on a Thursday evening. The challenge is conducted outside of class hours so some allowance is needed for students who cannot attend. The rules require at least one person from each group to be present. This has been successful. In addition, all groups are allowed to request a preferred time if there is a personal conflict. This option has been rarely exercised.

The challenge requires about 15–20 minutes per section and is dependent on the number of test sites available. With 13 sections, the University of Wyoming typically schedules an entire morning to complete the challenge.

Faculty and peer assistant participation during the challenge is needed. Typically, the course coordinator and the H. T. Person Chair are present for the entire challenge and serve as the adjudicators for safety and rule violations. The peer assistants handle the administrative tasks. Professors are encouraged to attend when their sections participate. Each challenge describes the peer assistant assignments. The peer assistant assignments are provided assuming 10–15 sections are participating in the challenge. A single section can be conducted with a faculty member and peer assistant.

4.2 FLEET EFFICIENCY AND THE AUTO DESIGN DILEMMA

Figure 4.1: Side impact safety test.

4.2.1 FACILITIES

This challenge requires a test track. Buildings with a concourse around the perimeter of the building work as does an indoor running track. The challenge is designed for a smooth floor. If the challenge is conducted on a running track with a composite surface, the qualifying distances may have to be adjusted to account for the higher ground friction.

A side impact test area needs to be established. Typically, this is an area about 12 feet square with a heavy plastic sheet placed on the floor. The impact hammer is placed in the center and the plastic collects the egg splatter. The challenge has been run in a smaller area with the impact hammer in a cardboard enclosure to capture the egg splash, but the visual impact of the test is greatly diminished.

4.2.2   THE CHALLENGE

This challenge calls for each section to form its own automobile company and to manufacture a fleet of electric motor powered automobiles that are both efficient and safe.

The EPA CAFE (Corporate Average Fuel Economy) rules require an increase in the average fuel efficiency for automobiles and trucks. Meeting these economy goals generally leads to a design solution favoring smaller engines and lighter weight cars. This creates a conflict between vehicle fuel efficiency and safety. Many lighter cars perform less well in collisions than heavier cars. Safety rules require side impact resistance and side airbags to help mitigate this problem.

In order to gain an insight into the design tradeoff between fuel efficiency and safety, each section will: a) construct a fleet of vehicles to meet fuel efficiency requirements and b) conduct side impact tests on vehicles that the team constructs. Fuel efficiency will be determined by a test of the distance traveled by your vehicles. The distance test begins with fresh batteries and runs until the vehicle stops. The impact test consists of a sledgehammer mounted on a frame that will deliver the side impact. The sledgehammer will be raised through a 90-degree angle. The vehicle will be placed against the side impact frame. An egg is placed in the driver/passenger compartment. The hammer is released to strike the car. The raw egg must survive.

The fleet consists of three sizes of vehicles: economy, standard, and SUV. All are powered by a 4.5 volt electric motor. Economy cars will use one AAA battery, standard cars use two AAA batteries, and SUVs use three AAA batteries, provided by the students. At least one vehicle of each size must be built and tested as described below. In that regard, the company (section) must decide what constitutes its best overall fleet composition.

The objective of each company is to earn the most points. The section with the highest point total gets the admiration of the entire freshman class and a pizza party on the last day of class.

Bonuses and Penalties will be assigned to each company as follows:

- Each vehicle meeting the minimum distance standard: 5 points.

- Bonus for fleet vehicles exceeding distance standards: 2 times the qualifying standard—10 points, 4 times—15 points, 8 times—20 points, etc. Each additional doubling of the qualifying standard gains 5 points.

- Companies not having at least one vehicle from each category will be penalized 50 points.

- Vehicles that fail to meet the minimum distance target: zero points.

- Each vehicle with an intact egg: 20 points.

- Each vehicle with a cracked egg: 5 points.

- Failure to pass the side impact test: zero points.

- All vehicles must be operated in a safe manner. Explosive, pyrotechnic, or similar devices are disqualified, and a 40 point deduction will be assessed to the team if they are brought to the challenge.

On the challenge day, vehicles will be weighed, measured, classified, and logged in prior to the test program. Heat times will be assigned in advance so each group knows when their fleet will run.

4.2.3 THE RULES

Companies consist of a complete section. Individual corporate divisions (groups) consist of two or three students. Single-person entries and teams greater than three students are not allowed. Remember, you are working as a company so internal communication is encouraged.

The sole source of power for the vehicle is a 4.5 volt electric motor provided in a kit to your group. Each kit includes the motor, gear set, and battery box (Figure 4.2).

Figure 4.2: Motor, battery box, and gears provided.
(Note: Jameco http://www.Jameco.com has motors and gears, pictured above, and provides torque/speed curves for the motors. Electric motors and gears are available from a number of web sites including Edmunds Scientific.)

Each team must maintain a design notebook (see Chapter 2).

No team shall spend more than $15.00 for supplies and equipment to manufacture the vehicle. It is OK to use free stuff. The definition of "free" is that it has no commercial value. In short, the instructor can elect to keep it or throw it in the trash following the competition. Modified prefabricated cars are automatically disqualified. "Rented" or "borrowed" materials are not allowed. This budget can be adjusted if motors and gears are not provided.

Vehicles will be weighed, measured, and logged in prior to the test program.

At your appointed time, take your vehicle to the distance trial station. Upon completion of the distance trials, remove the batteries and place them in the recycling box. Take your vehicle to the Safety Test station.

Vehicle fabrication:

- The vehicle is to be constructed using only 1/4-inch-thick foamcore board, engineering calculation paper, and white (Elmer's) or thermo plastic (hot-melt) glue. The vehicle must fit within an envelope that is 10 inches long, 4 inches wide, and 3 inches tall. Place a mark on the centerline of the driver location on your vehicle. This corresponds to approximately 1:18 scale of an actual car.

- A box having these inside dimensions will serve as a template to verify compliance of the vehicle size. The floor of the vehicle must be at least 1/2 inch above the ground level (Figure 4.3) under impact criteria. A 1/2-inch-thick block must pass freely beneath the vehicle.

- Wheels, motor mounts, axles, and gear shafts may be metal or plastic as appropriate. The vehicle must be hollow; however, you can consider the use of transverse elements for the firewall and floor bracing.

- You may consider stiffening elements at the floor, doors, and doorposts, and strategic placement of the battery box. No seats or other accoutrements are required; however the "sheet metal" parts of the cars should not be so flimsy that the aesthetics of the car suffers.

- The car body should be shaped to include the passenger compartment, hood, and trunk area. Front and rear windshield areas must be open. Doors need not open, but there must be a way to replace batteries without damaging the vehicle.

- There must be a hole 1 3/4 inches in diameter on one side or the roof for the insertion of the egg. You must decide what portions of the car may be fabricated from sheet paper and where the foamcore may be used as strengthening elements.

Vehicle weights with batteries but without the egg are as follows:

- Compact car (1 battery): less than 225 grams

- Standard car (2 batteries): 225 to 300 grams

- SUV (3 batteries): greater than 300 grams

The distance criteria:

- The vehicle and driver (the egg) will be placed at the starting line. The vehicle is started by turning on the power. No push starts are allowed. The total distance traveled around the track will be measured. Teams may redirect the cars to keep them in their lanes.

- To qualify, a vehicle must travel a minimum of 100 yards.

- Masking tape lines on the floor will indicate the qualifying distance and bonus point locations.

The vehicle safety criteria test:

- Side impact safety tests are conducted with the apparatus shown in Figure 4.3.

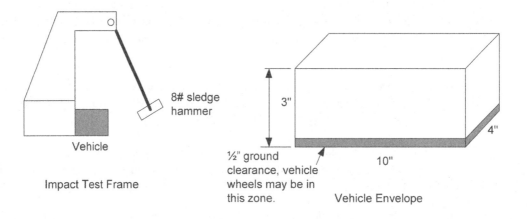

Figure 4.3: Side impact safety test apparatus and vehicle envelope.

Once each car has completed its distance trial, the batteries are removed, placed in the recycle bin, and the vehicle is taken to the impact test machine.

- The egg is reinstalled and the car placed in the impact testing machine with the centerline aligned with the hammer. The egg is placed loosely in the vehicle. "Seatbelts" and "airbags" are permitted if they are in the car during the distance trial. The hammer handle is moved to the horizontal position and then released.

- An egg is deemed to have survived if there are no cracks in the shell. Therefore, careful removal of the egg after the impact test is important.

4.2.4   SAFETY

The objective of the challenge is to foster engineering creativity and cooperation. The design group is responsible for ensuring safety of participants and spectators during the challenge. Groups using any feature deemed dangerous by the judges may be asked at any time to prepare a safety plan or

(a) Test setup   (b) Results

Figure 4.4: Side impact test execution.

suitably modify the design before continuing in the challenge. Offending designs may be disqualified at the discretion of the faculty member or peer assistants. Use of pyrotechnic or similar devices is strictly prohibited. Any questions regarding safety may be directed to your instructor.

4.2.5   CHALLENGE ORGANIZATION

Early organization:

- Identify a test facility track and schedule the design challenge day
- Identify a practice area and setup.

Preparation for challenge day:

- Prepare an overall summary score spreadsheet
- Prepare data sheets and coordinate individual score sheets with the summary spreadsheet
- Prepare compliance box
- Build the impact test hammer
- Prepare a recycling box for used batteries
- Draw lots for the section challenge times
- Prepare a press release if appropriate

- Prepare a descriptive poster if the challenge is in a public area

Challenge day: Peer assistant assignments:

- Set up a registration table: log in each group, weigh and measure the vehicle, score the design notebooks, and provide the individual data sheet: Typically, four assistants

- Distance qualifications: Typically, two assistants. The assistants lay out the qualification and bonus lines and confirm the distance traveled.

- Side impact test: Typically, two assistants. The assistants confirm the outcome, sign off on the data sheet, and assure each group complete their cleanup.

- Data recording: Students turn in data sheets and one or two assistants enter the data into the summary spreadsheet. One person may have to go back to the other activities to confirm data, so two people are preferred.

- Oversee the site clean-up: all.

Following the challenge:

- Provide the summary data sheet to all instructors.

- Arrange any awards, e.g., pizza party for the best section.

4.2.6 CONCLUDING COMMENTS

This challenge has been very successful. About 80 percent of the vehicles make the distance requirement and approximately 25 percent of the vehicles pass the entire challenge. One vehicle traveled nearly 1,000 yards. As seen in the photos, the side impact test is extremely popular. Aligning the hammer to the level position, so there is no excess energy and no "cutting corners," is a valuable experience, especially with the competitors watching to keep everything fair.

4.3 DUNEBUGGY DASH

4.3.1 FACILITIES

This challenge requires an obstacle course. As seen below, the challenge was completed in the Civil Engineering Sediment Transport facility lab. A dirt road, grassy strip, gravel pile, or similar terrain is acceptable. The challenge should require the students to consider the effects of foreign materials getting into critical parts of the vehicle and affecting performance. The challenge is modeled after the Mars Rover, so lightweight motors are emphasized.

4.3.2 CHALLENGE

This challenge requires construction of an electric motor powered dune buggy that can drive across the sediment transportation laboratory "sandbox" shown below in Figure 4.5.

(a) Spring

(b) Fall

Figure 4.5: Challenge course.

4.3.3 THE RULES

- Teams consist of three or four students. Single and two-person entries are not allowed. This size limitation is partially based on the size of the lab where the challenge will take place. If space is available, two- and three-person teams are preferred.

- The sole source of power for the vehicle is a 4.5 volt electric motor available from a kit in the Dean's office. The price is $5.00 and includes the motor, gear set, and battery box (Figure 4.2). (Note: The motors provided in the kit are underpowered and require the students to adjust the gear ratios for successful results. The kits cost more than $5.00 so the project is partially subsidized. The motors are required otherwise students will use commercial toy dune buggy motors, which are far more effective.)

- Each team must maintain a design notebook (see Chapter 2).

- No team shall spend more than $20.00 for supplies and equipment to manufacture the dune buggy. This includes the $5.00 for the motor so each group has a $15.00 operating budget for other materials. It is OK to use free stuff. The definition of "free" is that it has no commercial value. In short, the instructor can elect to keep it or throw it in the trash following the competition. "Rented" or "borrowed" materials are not allowed.

- The vehicle will be placed on the sand with the back of the buggy touching the south wall. The vehicle is started by turning on the power. No push starts are allowed.

- No one is allowed on the sand. A peer assistant is on a movable walkway to collect stalled or stranded vehicles.

- All vehicles must be operated in a safe manner.

Recognition will be given for the following categories:

- Rube Goldberg Award–most complicated design that actually works.

- Students' Choice Award.

- Instructor and Peer Assistant Award in each section.

- The section with the highest average distance travelled by all vehicles in the section receives a pizza party.

4.3.4   CHALLENGE ORGANIZATION

Early Organization:

- Identify a test facility and schedule the design challenge day.

- Identify a practice area and setup.

Preparation for challenge day:

- Prepare summary score spreadsheet.

- Prepare data sheets and coordinate individual score sheets with the summary spreadsheet.

- Prepare a recycling box for used batteries.

- Draw lots for the section challenge times.

- Prepare a press release if appropriate.

- Prepare a descriptive poster if the challenge is in a public area.

Challenge day: Peer assistant assignments

- Set up a registration table: log in each group, verify the motors, and provide the individual data sheet: Typically, two assistants.

- Distance qualifications: Typically, two assistants. The assistants confirm the distance traversed.

- Recovery team: Typically, two assistants. The assistants recover stalled or damaged vehicles.

- Data recording: One or two assistants enter the data into the summary spreadsheet. One person may have to go back to the other activities to confirm data, so two people are preferred.

- Oversee the site clean-up: all.

Following the challenge:

- Provide the summary data sheet to all instructors.

- Arrange any awards, e.g., pizza party or gift certificates for best categories.

4.3.5   CONCLUDING COMMENTS

An underlying assumption in this challenge was for the students to experience the difficulty of working in "dirty" environments. The sand in the sediment basin was ideal for this. Background for this challenge included discussion of the Mars Rovers.

About half the vehicles bogged down before moving five feet. Gearing was also critical since the motors were high rpm and low torque. Some groups elected to use balloons to lift the vehicle and the motor for a propeller to move it across the test area. Thus, the challenge was changed from "traverse" the facility to "drive across" the facility. One group built a catapult to propel the vehicle across. After almost hitting a student on the opposite side, the design was judged "unsafe." Lastly, recognition took the form of in-class awards for the different categories. The instructors and peer assistants made comments on the evaluation sheets and these formed the basis of the selections.

4.4   FIRST FLIGHT: FLY A FOAM AIRPLANE 100 YARDS

Figure 4.6: Foam airplane flight preparation.

4.4.1   FACILITIES

This project requires a large open field. We selected the indoor football practice facility for two reasons. First, by October, Wyoming weather conditions can be snowy and an outside activity may be compromised. Second, the average wind in Laramie in October is in the 10–30 mph range and is highly variable. This creates a potentially unfair comparison condition. By moving inside we provided a stable environment that comes with a very nice graduated scale on the floor.

4.4.2   CHALLENGE

This challenge is to modify a foam plane to fly 100 yards—the length of a football field. The challenge will be held in the football indoor practice facility.

> *From **Wikipedia**: the definition of an airplane: A **fixed wing aircraft** is an aircraft capable of flight using wings that generate lift due to the vehicle's forward airspeed and the shape of the wings.*

Each section will function as a team. A team consists of groups of two to three students. Each group will receive one foam model plane. The basic plane is to be redesigned, modified, and tested to optimize the number of planes in the team to make the required distance. In that regard, the team must decide what constitutes its best selection of power and flight strategies. Individual groups present their ideas in class and receive input from the team.

- Each group works to design a plane to optimize the team response in the challenge.

- A successful plane will fly 100 yards from end zone to end zone.

- Each group must maintain a design notebook (see Chapter 2).

- One plane is supplied to each group and is classified as "free."

- No group shall spend more than $30.00 for supplies to equipment to modify the plane. It is OK to use "free stuff." The definition of "free stuff" is that it has no commercial value. In short, the instructor can elect to keep it or throw it away following the challenge. "Rented" or "borrowed" materials are not allowed. Only the cost of the final plane need be included in the budget. If a plane is destroyed in testing, the cost of a new plane must be included in the budget.

- All planes must be designed, fabricated, and tested prior to submittal.

4.4.3   THE RULES

- The plane may be powered by any safe device including elastic bands, launchers, electric motors, or other mechanical contraptions. *For safety reasons, no gasoline or model airplane engines or rocket engines (e.g., Estes) are allowed.* Part of the challenge is for the team to optimize flight selection and design.

- The plane must be launched behind the goal line and attempt to cross the opposite goal line. As in football, if just the nose crosses the line, it is a success. Length is measured to the final position of the nose of the aircraft. (Bounces on the ground count.)

- All components of the original plane must be used in the challenge; however, decals and tickers are optional.

- Following initial testing, the group may elect to construct a new plane based on the performance of the trial runs.

- By definition, planes fly by lift. They are not dragged, towed, pulled, fastened to a wire, or suspended from balloons.

- On the challenge day, the team will have ten minutes to fly the length of the field.

Scoring

Each team may fly as many planes as there are groups. The team score is the average distance traversed by the best flight of each group. At least three groups must fly to qualify. Thus, if six planes fly the full length of the field and two fly 20 yards, the score is 80 points (6 flights x 100 yards + 2 x 20)/8. In addition, each team receives and additional points based on the design notebook grade.

Awards

The team with the highest overall score will receive a pizza party during the final class session.

4.4.4   SAFETY

The objective of the challenge is to foster engineering creativity and cooperation. The design group is responsible for ensuring safety of participants and spectators during the challenge. Groups using any feature deemed dangerous by the judges may be asked at any time to prepare a safety plan or suitably modify the design before continuing in the challenge. Offending designs may be disqualified at the discretion of the faculty member or peer assistants. Use of pyrotechnic or similar devices is strictly prohibited. Any questions regarding safety may be directed to your instructor.

4.4.5   CHALLENGE ORGANIZATION

Early Organization:

- Identify a test facility track and schedule the design challenge day.

- Identify a practice area and setup. This is often the same facility and scheduling the facility for two activities is a critical endeavor.

Preparation for challenge day:

- Prepare summary score spreadsheet.

- Prepare data sheets and coordinate individual score sheets with the summary spreadsheet.

- Prepare a recycling box for used batteries.

- Draw lots for the section challenge times.

- Prepare a press release if appropriate.

- Prepare a descriptive poster if the challenge is in a public area.

Challenge day: Peer assistant assignments:

- Set up a registration table: log in each group, inspect for safety issues, grade notebooks, and provide the individual data sheet: Typically, two assistants.

- Distance qualifications: Typically, two assistants. The assistants confirm the distance flown.

- Data recording: Two or three assistants verify the data on the summary spreadsheet. One person may have to go back to the other activities to confirm data, so two people are preferred.

- Oversee site clean-up: all.

Following the challenge:

- Provide the summary data sheet to all instructors.

- Arrange award, e.g., pizza party or gift certificates for best categories.

4.4.6   CONCLUDING COMMENTS

Figure 4.7: Plane launch.

Class lectures addressed power, lift, speed, drag, and launch options. The longest flight was just over 60 yards. The principle difficulty was a high speed launch raised the nose of the plane to a stall position followed by the plane crashing.

There were three weeks between the trial and the final design. Most students used this time to improve their designs. Some of the planes crashed during testing and students did not rebuild them for the final challenge. This again is a function of the scoring and emphasis on participation not success. At least one group used CO_2 cartridges for propulsion. The group was required to present a safety plan to assure that the cartridges were secured and controlled when fired.

Two groups tried to tow their planes across the field. They were disqualified. Several groups tried using balloons to lift the plane. The drag from the balloons generally resulted in the plane flying in circles. The definition of an airplane and the requirement for the plan to fly by its own lift should limit balloons in the challenge.

This challenge included the value of the design notebooks in the overall scoring. The inclusion was to place additional emphasis on the value of recording progress.

4.5   HACKYSACK FLIP

4.5.1   FACILITIES

The number of available tracks is a limiting feature for planning the amount of time needed to complete the challenge. Three tracks were constructed for this challenge. The challenge was held on the stage of one of the auditoriums on campus. This provided a waiting and viewing area for students and guests. This challenge is well suited for public viewing.

4.5.2   CHALLENGE

This challenge requires each section to form its own company to manufacture a fleet of electric motor powered vehicles that runs on a prefabricated track and can perform the tasks listed below. Each section will enter 8–10 vehicles. Scoring will be completed by averaging the total number of points from each vehicle's best run.

Each section will function as a team. A team consists of groups of two to three students. Vehicles are to be designed, built, and tested to optimize the team's point score. In that regard, the team must decide what constitutes its best composition of vehicles. Individual groups present their ideas and receive input from the team.

Each group designs and constructs a vehicle capable of performing the following tasks:

1. *Pass* through the start gate (S),

2. *ascend* a 20-degree hill,

3. *propel* a "hackysack" bean bag through the hole in a vertical wall (W) at the top of the hill,

4. *knock down* a flag or flags located at the top of the hill, and

Figure 4.8: Challenge day setup.

5. *descend the hill* and round the exit run-out to the finish line.

The points scored on the best run will be recorded.

The Hackysacks
Details of the Hackysacks are given in Figure 4.9. They weigh 37 to 39.8 grams and are approximately 2 1/2 in. in diameter.

The Track
Figure 4.10 shows the approximate dimensions of the track. The drawing is not to scale. The track width dimension may vary by ± 0.5 inches at any point. The side rails are made from 2.5-inch tall pressboard. The carpet is a standard, commercial grade. Contestants will approach the right-hand side of the ramp as seen from the side view (see S below).

The "top of the hill" zone is defined by the two lines (T). A dowel extends 1" into the track opposite the "Hackysack" wall (W). The Hackysack must be launched through the hole in the wall. The diameter of the hacky sack hole is 5." The flags, each consisting of a dowel extending

Figure 4.9: Hackysack details.

approximately 1.4 inches above the track wall (see illustration), are mounted on either side of the track at the centerline. You may lower either or both flags as you exit the top of the hill. A vehicle's flag will pivot only in the direction of the forward motion of the vehicle. Design should include consideration of "high centering" at the start and top of the hill climb. One track is available for testing cars prior to the challenge.

4.5.3 THE RULES

Vehicle Design Specifications

1. The complete vehicle must be designed to fit inside a 6-inch cube. The complete vehicle is defined by all its parts. Appendages, such as an arm, may extend beyond this limit once activated by passing through the vehicle portal, P, but cannot be activated before the start of the run.

2. The vehicle must remain intact throughout the competition, that is, it may not jettison any unattached part, and may not divide into two or more separate sections or pieces. All parts must remain attached to the vehicle. For the purpose of this rule, the definition of "attached" is meant to exclude attachment by string, wire, or other flexible tether.

3. The weight of the vehicle, including batteries, must not exceed 1.0 kg (2.2 pounds).

4. Peer assistants will supply competition "Hackysacks" at each ramp. "Hackysack" specifications are attached in Figure 2.

5. The vehicle must be stationary prior to the start, and it cannot be pushed by a group member as part of the start. After the start signal, the vehicle's propulsion system may be activated using the switch on the battery box, but cannot be activated prior to the start.

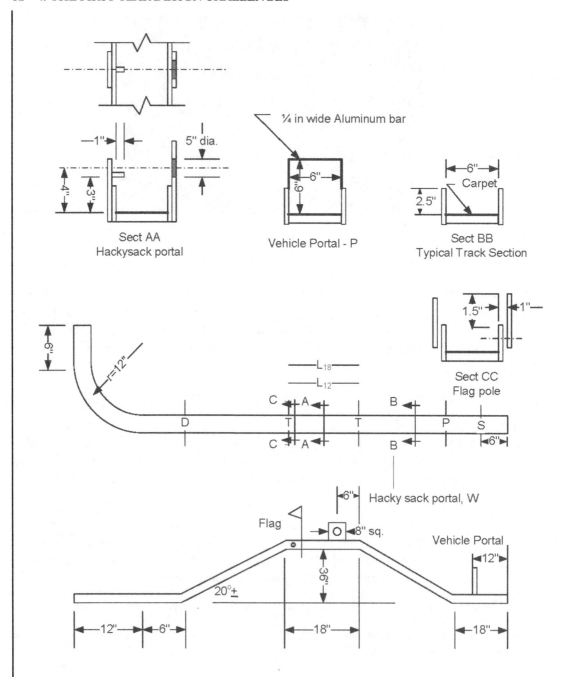

Figure 4.10: Schematic test track.

6. In the execution of its tasks, the vehicle may not damage the track, its walls, or the roadway carpet.

7. Onboard computing devices, such as microcontrollers, are not permitted.

Power
Power to propel the vehicle and to run any onboard activation or electronic devices is derived using two AAA 1.5 V batteries and one 1.5–4.5 volt DC motor supplied in the parts kit (Figure 4.2). Batteries may be connected to the vehicle in any configuration. Supplemental mechanical power may be derived from such devices as springs, mousetraps, balloons, and rubber bands. Compressed gas cylinders, chemical reactions, or combustion of any type are *not* allowed. Mercury switches of any type are *not* allowed.

Challenge Day Procedures
Team members will have two minutes to reach the starting position after being called for a round. At the judges' discretion, any vehicle not ready after the two minute countdown will forfeit the round and may be allowed to compete after all other teams have completed their rounds.

A timer will count down the five minutes for each round. Groups can have as many runs as possible within the five minutes. The run will last until the vehicle crosses the finish line. Contestants may maintain contact with vehicles prior to "go" but may not touch vehicles during the run interval. If a group member touches its vehicle before run has been completed, the run will be considered incomplete, and the accumulated score to that point will be reduced by two points.

All vehicle wheels must remain within the side rails of the track. Deployed appendages may extend beyond the side rails after the start, but the tops of the side rails may not be used to support the vehicle.

Groups may provide a fresh set of batteries at the initiation of their round. The round must be completed on the fresh batteries. If the batteries die, points accumulated up to stop of the vehicle will be counted.

Groups may modify vehicles between runs; however, the time limit for the round will be maintained.

Only the called groups may enter the stage competition area and the vehicle repair area. Similarly, only team members may request verification of opposing vehicles for compliance with contest rules and design limits. Spectators are not permitted to make such requests.

Any vehicle compliance challenge must be made to the judges prior to the awarding of points for a particular round. If a vehicle is challenged and found to violate context requirements, it may be disqualified.

Scoring challenges for a particular round must be made by the end of that round of the competition and may only be made by team members. Resolution of point challenges will be made at the sole discretion of the judges. Vehicle modification and rerun following disqualification is at the sole discretion of the professor.

Scoring

For each run of the round, a maximum of 9 points will be awarded for each run as follows:

Ascending Points: One point will be awarded to each vehicle that successfully ascends to the "top of the hill" of the ramp. To earn "top of the hill" points, the vehicle, and all its parts, must reside between points (T).

Hackysack Points: One point will be awarded to each vehicle that successfully propels its hackysack through the hole in the wall. An additional point will be awarded to each vehicle that propels the entire diameter of its hackysack beyond the (L_1) line and two points for the hackysack completely passing the L_2 line.

Descending Points: One point will be awarded to each vehicle that passes, including all its parts, the bottom of the descending ramp (D). To receive descending points, a vehicle must first earn ascending points.

Flag Points: One point will be awarded to each flag knocked down during the run.

Run-out Points: Two points will be awarded to any vehicle that completely exits the run-out curve. Completely exit means that a ruler may be placed between the end of the track and the end of the vehicle.

Penalty Deductions: Two points will be deducted from the run for a vehicle not completing the course. The minimum score for a run is zero; penalty points cannot yield negative scores.

Vehicle Mishap: If a vehicle falls off the ramp for any reason or the batteries die, it will retain points earned for the run prior to the mishap.

Points for the best run are recorded.

General Rules

- Each group must maintain a design notebook (see Chapter 2).

- No group shall spend more than $15.00 for supplies and equipment to manufacture its vehicle. It is OK to use "free stuff." The definition of "free stuff" is that it has no commercial value. In short, the instructor can elect to keep it or throw it in the trash following the competition. Modified prefabricated cars are automatically disqualified. "Rented" or "borrowed" materials are not allowed.

- Vehicles will be weighed, measured, and logged in prior to the test program.

- Vehicle fabrication: The vehicles must be constructed from scratch, that is, no premade plastic bodies can be used. While the choice of materials is left to each group, 1/2 inch thick foamcore board, white (Elmer's) or thermo plastic (hot-melt) glue have proven to be a suitable construction material, except for axles and wheels, and are available at the campus store.

Prior to the start of the challenge, groups must conduct a "calibration round" on the test track and record the results in the design notebook.

No tools will be supplied on the competition day. teams are expected to bring all necessary items to repair or modify vehicles during the competition, including spare parts.

After the official start of the competition, only registered student contestants will be allowed in the competition and work areas. Spectators are welcome to view the competition from the seating area in the auditorium.

After each run, and prior to leaving the ramp area, groups are responsible for verifying that point totals have been correctly recorded by the ramp judge and that the challenge area is clean. Judges are instructed to oversee these checks. One group member will sign each score sheet.

4.5.4   SAFETY

The objective of the challenge is to foster engineering creativity and cooperation. The design group is responsible for ensuring safety of participants and spectators during the challenge. Groups using any feature deemed dangerous by the judges may be asked at any time to prepare a safety plan or suitably modify the design before continuing in the challenge. Offending designs may be disqualified at the discretion of the faculty member or peer assistants. Use of pyrotechnic or similar devices is strictly prohibited. Any questions regarding safety may be directed to your instructor.

Students have access to the College of Engineering and Applied Science shops. Sign up for shop time is required and students must have watched the safety practices video presented in class.

4.5.5   CHALLENGE ORGANIZATION

Early Organization:

- Identify a challenge facility area and schedule the design challenge day.

- Fabricate the test tracks.

- Identify a practice area and set up one track for trial runs.

Preparation for challenge day:

- Prepare summary score spreadsheet.

- Prepare data sheets and coordinate individual score sheets with the summary spreadsheet.

- Prepare a recycling box for used batteries.

- Draw lots for the section challenge times.

- Prepare a press release if appropriate.

- Prepare a descriptive poster if the challenge is in a public area.

Challenge day: Peer assistant assignments:

- Move the challenge tracks into place before students arrive.

- Set up a registration table: log in each group, measure vehicles, grade notebooks, and provide the individual data sheet: Typically, two assistants.

- Run qualifications: Typically, three assistants, one per track. The assistants confirm the points earned.

- Data recording: One or two assistants enter the data into the summary spreadsheet. One person may have to go back to the other activities to confirm data, so two people are preferred.

- Oversee site clean-up: all.

Following the challenge:

- Provide the summary data sheet to all instructors.

- Arrange any awards, e.g., pizza party.

4.5.6   CONCLUDING COMMENTS

Figure 4.11: Vehicle testing.

This was the most complex challenge we attempted because of limits imposed by the test track. The college shop fabricated the tracks. One track was set up in a lab space for students to test their designs. The few days prior to the challenge the room was full.

About 10% of the vehicles completed the challenge. The original challenge has a 30° ramp and many vehicles had great difficulty making the climb. This challenge has been reduced to 20° to improve the opportunity for success. Several cars rolled over on the descending ramp. The lower grade should alleviate this problem. The end runout also proved to be a stumbling block as cars were getting stuck. An 18 in. radius runout would resolve this issue.

4.6 MOUSETRAP POWERED CAR SLALOM

4.6.1 FACILITIES

A hard floor 25 feet x 25 feet works best and allows the slalom course to be set up and the width allows 6 to 8 lanes to be laid out side by side. Tight carpet also works but the decision should be made prior to initiation of the challenge as it will affect both distance and steering. A meeting room in the student union or equivalent provides high visibility and public access.

4.6.2 CHALLENGE

This challenge requires each group to construct a mousetrap-powered car that can negotiate a slalom course consisting of four 3 3/4 in square pylons placed in a straight line 4 foot on centers.

4.6.3 THE RULES

Each section will function as a team. A team consists of groups of two to three students. Cars are to be designed, built, and tested to optimize the team's point score. In that regard, the team must decide what constitutes its best composition of cars.

Individual groups present their ideas in class and receive input from the team.

- The team works to design vehicles to optimize the team response in the challenge.

- Each group must design and fabricate at least one car.

- Mousetraps will be supplied in class.

- In addition to the two mousetraps supplied, no group shall spend more than $20.00 for supplies and equipment to manufacture its car. It is OK to use "free stuff." The definition of "free stuff" is that it has no commercial value. In short, the instructor can elect to keep it or throw it in the trash following the challenge. "Rented" or "borrowed" materials are not allowed.

- All cars must be designed, fabricated, and tested prior to submittal.

- Groups will consist of two or three students. Single-person entries are not allowed.

- The sole source of power for the vehicle is a mousetrap; rattraps are not allowed.

- The vehicle may use one or two mousetraps in any combination for power and/or steering.

- Each group must maintain a design notebook (see Chapter 2).

- The vehicle will be placed at the start line and released by the group members. Each vehicle must be placed with its longitudinal axis parallel to the course. Wheels may be aligned at the group's discretion (see layout in Figure 4.12). Once released, the car cannot be touched by any group member. The vehicle must cross the centerline before passing the first pylon (see Figure 4.12).

- Prior to the actual challenge, each car should complete a successful trial run and distances of each run should be recorded in the design notebook.

- On the challenge day, the group will have five minutes to make a successful run.

- Vehicle fabrication: The car must be constructed from scratch, that is, no premade units, e.g., premade steering mechanisms. The choice of materials is left to each group.

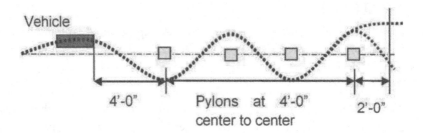

Figure 4.12: Challenge course and starting configuration.

Scoring

Each group receives 0 to 4 points based on the number of pylons successfully passed. The group earns an additional 2 points for completing the run and crossing the finish line. The team score is the average of the group scores.

Developing a Test Program

Engineering design and development requires validation and testing. Your notebook must clearly identify a development schedule, test dates, and test results. Modification to the design after each test must be documented. The best run and points must be clearly recorded and dated in the notebook.

4.6.4   SAFETY

The objective of the challenge is to foster engineering creativity and cooperation. The design groups are responsible for ensuring safety of participants and spectators during the challenge. Groups using any feature deemed dangerous by the judges may be asked at any time to suitably modify the design before continuing in the challenge. Offending vehicles may be disqualified at the discretion of the professor. Use of pyrotechnic or similar devices is strictly prohibited.

4.6.5   CHALLENGE ORGANIZATION

Early Organization:

- Identify a challenge facility track and schedule the design challenge day.

- Identify a practice area and setup. The initial practice area can be a single track in a classroom.

Preparation for challenge day:

- Prepare summary score spreadsheet.

- Prepare data sheets and coordinate individual score sheets with the summary spreadsheet.

- Fabricate pylons.

- Draw lots for the section challenge times.

- Prepare a press release if appropriate.

- Prepare a descriptive poster if the challenge is in a public area.

Challenge day: Peer assistant assignments:

- Set up pylons, start and finish lines prior to the students arriving.

- Set up a registration table: log in each group, verify the power source, grade notebooks, and provide the individual data sheet: Typically, two assistants

- Run qualifications: Typically, four assistants. The assistants confirm the course was successfully traversed.

- Data recording: One or two assistants enter the number of points into the summary spreadsheet. One person may have to go back to the other activities to confirm data, so two people are preferred.

- Oversee site clean-up: all.

Following the challenge:

- Provide the summary data sheet to all instructors.

- Arrange any awards, e.g., pizza party or gift certificates for best categories.

4.6.6   CONCLUDING COMMENTS

This challenge is deceptive. Many students have participated in mousetrap cars in high school. The autonomous steering, however, requires closer coordination of forward progress and sinusoidal action. At least two groups in the challenge used mousetrap-powered cars and radio-controlled steering. One student group brought in a RC controller with the comment that sabotage is not disallowed! The rule eliminating prefabricated steering is intended to exclude the most egregious use of RC controls.

This challenge needed a gross of mousetraps. No one in town stocked that many so they were special ordered and provided to the teams. This can be a professional student group fund raiser.

4.7   THE GREAT WALL OF CARPET

Figure 4.13: Carpet climb challenge.

4.7.1 FACILITIES

This challenge works best with a two-story atrium and carpet hung from the upper floor. The particular challenge was held in the atrium of the campus library. Two similar types of carpet were used. The carpet strips were approximately 2 feet wide. The carpets are hung so only the fabric side is available for the challenge.

4.7.2 CHALLENGE

This challenge requires construction of an autonomous robotic device that can climb a carpet draped in the atrium of a two-story building. Each group will design and construct a robot according to the rules laid out below. To qualify, a robot must climb at least two vertical feet.

4.7.3 THE RULES

Each section will function as a team. A team consists of groups of two to three students. Robots are to be designed, built, and tested to optimize the team's point score. In that regard, the team must decide what constitutes its best composition of robots.

Individual groups present their ideas in class and receive input from the team.

- The team works to design vehicles to optimize the team response in the challenge.

- Each group must design and fabricate one robot.

- Each group must maintain a design notebook (see Chapter 2).

- No group shall spend more than $40.00 for supplies and equipment to manufacture its robot. It is OK to use "free stuff." The definition of "free" is that it has no commercial value. In short, the instructor can elect to keep it or throw it in the trash following the competition. "Rented" or "borrowed" materials are not allowed.

- All robots must be designed, fabricated, and tested prior to submittal.

- The robot must be constructed from scratch, that is, no premade units can be used. The choice of materials is left to each group.

- Groups will consist of two or three students. Single person entries and groups of more than three students are not allowed.

- The robot may be powered by any device, elastic bands, electric motors, or other mechanical contraptions. Part of the challenge is for the team to optimize very small lightweight climbers with heavier and more powerful alternatives.

- The entire robot must make the climb; however the robot may work in discrete elements if needed and the robot must grip the carpet to climb.

- The robot will be placed at the start line and released by the group members. Each robot must function autonomously, that is, it cannot be touched by a group member while operating. If touched, it must be restarted.

- Prior to the actual challenge, each robot should complete a successful qualifying run. A survey rod will be provided and the height for each run recorded.

- On the challenge day, the robot will have five minutes to make a successful run.

- **Special note: there may be two different carpet types. You are not assured which one you will get on the challenge day.**

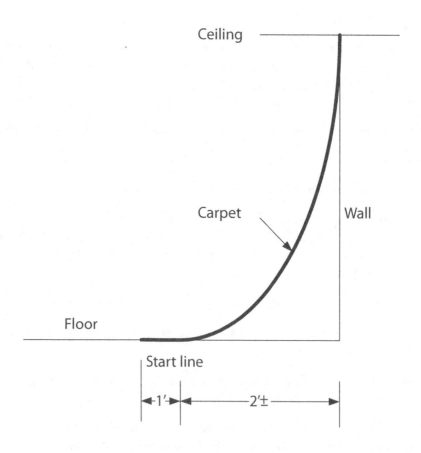

Figure 4.14: Approximate challenge course and starting configuration.

Developing a Test Program

Engineering design and development requires validation and testing. Your notebook must clearly identify a development schedule, test dates, and test results. Modification to the design after each test must be documented. The runs and climb height must be clearly recorded and dated in the design notebook.

4.7.4 SCORING

The best height of the challenge run will be recorded. If the robot does not climb in the challenge, a challenge height of zero will be recorded. The team with the highest average climb height divided by the average cost will be recognized as the highest climbing/most efficient team on campus!

To start each run, the judge will indicate start, and the timing clock will begin. The test will last until the run is complete or time expires. Groups may maintain contact with robots prior to the robot start but may not touch the robot during the run. A group member may touch the robot to prevent damage but the run must then be restarted. The run record must be completed without manual assistance or the run will be disqualified.

Only team members may request verification of opposing robots for compliance with contest rules and design limits. Spectators are not permitted to make such requests.

Any challenge must be made to the judges prior to the awarding of points for a particular test. If a robot is challenged and found to violate context requirements, it will be disqualified and a height of zero recorded.

4.7.5 SOME REFERENCES

If you have never seen a climbing robot, consider looking on Bing or Google. There are a number of good sites, many of which are far more complex than needed for this challenge. The description of these robots will give you an idea of some of the design considerations to be included in your project.

4.7.6 SAFETY

The objective of the challenge is to foster engineering creativity and cooperation. The design group is responsible for ensuring safety of participants and spectators during the challenge. Groups using any feature deemed dangerous by the judges may be asked at any time to prepare a safety plan or suitably modify the design before continuing in the challenge. Offending designs may be disqualified at the discretion of the faculty member or peer assistants. Use of pyrotechnic or similar devices is strictly prohibited. Any questions regarding safety may be directed to your instructor.

4.7.7 CHALLENGE ORGANIZATION

Early Organization:

- Identify a challenge facility track and schedule the design challenge day.

- Identify a practice area and setup one of each type of carpet.

Preparation for challenge day:

- Procure the challenge carpet. As the rules note, we had two different types of carpet so the students did not know which carpet they will climb until the challenge.

- Prepare summary score spreadsheet.

- Prepare data sheets and coordinate individual score sheets with the summary spreadsheet.

- Draw lots for the section challenge times.

- Prepare a press release if appropriate.

- Prepare a descriptive poster if the challenge is in a public area.

Challenge day: Peer assistant assignments:

- Hang carpet and arrange survey rods.

- Set up a registration table: log in each group, grade notebooks, and provide the individual data sheet: Typically, two assistants.

- Distance qualifications: four assistants. The assistants confirm the distance was climbed. We set up four climbing stations and brought over survey rods to assist with measuring the height climbed. Place masking tape at the challenge finish.

- Data recording: One or two assistants enter the data into the summary spreadsheet. One person may have to go back to the other activities to confirm data, so two people are preferred.

- Oversee site clean-up: all.

Following the challenge:

- Provide the summary data sheet to all instructors.

- Arrange any awards, e.g., pizza party.

4.7.8   CONCLUDING COMMENTS

Only two of 131 climbers made it to the top and one design used helium filled balloons to lift the robot. These results made this the most difficult challenge in terms of success. Nothing in the rules precluded using the edge of the carpet, and several groups used this option. One group made a catapult with a three-pronged fishing hook. The design launched the hook then used a winch to pull itself up. It performed poorly and required a safety plan for throwing the hook. About 30 years ago the University of Utah had a similar challenge using small elastic band-powered paperclip climbers.

Figure 4.15: Edge climber.

4.8   UNDERWATER RECOVERY VEHICLE

4.8.1   FACILITIES

The challenge requires use of the university swimming pool. In addition to gaining access to the pool, life guards are hired to assure safety. No students are allowed in the pool, hence the requirement for the robots to be tethered. Local pool rules and depths must be incorporated in the rules.

4.8.2   CHALLENGE

This challenge is to construct a tethered robotic device that can retrieve a marker from the bottom of the swimming pool according to the rules laid out below. The final challenge will be in the shallow end of the pool and the pool will not available for practice prior to the challenge. A stock watering tank will be set up and available for practice.

Figure 4.16: The beginning of the underwater challenge.

4.8.3 THE RULES

Each section will function as a team. A team consists of groups of two to three students. Robots are to be designed, built, and tested to optimize the number of retrieved weights. In that regard, the team must decide what constitutes its best selection of robots; however, each team must have at least two underwater robots (submarines) and two surface robots.

Individual groups present their ideas in class and receive input from the team.

- The team works to design vehicles to optimize the team response in the challenge.

- Each group must design and fabricate one robot.

- A successful robot will retrieve a designated small weight from the bottom of the campus pool by picking up the hook on the weight (Figure 4.13). A steel washer will be placed on the magnet for the challenge. (Note: the washer prevents the magnetic base from attaching to a pool drain.)

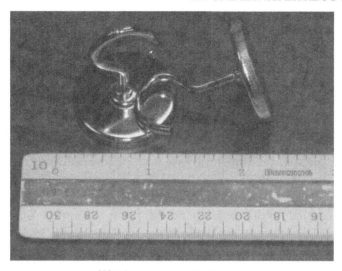

Weight approximately 30 grams

Figure 4.17: Marker to be retrieved from pool.

- Each group must maintain a design notebook (see Chapter 2).

- No group shall spend more than $50.00 for supplies and equipment to manufacture its robot. It is OK to use "free stuff." The definition of "free stuff" is that it has no commercial value. In short, the instructor can elect to keep it or throw it in the trash following the challenge. "Rented" or "borrowed" materials are not allowed.

- All robots must be designed, fabricated, and tested prior to submittal

- Robot fabrication: The robot must be constructed from scratch, that is, no premade boats may be used, but premade components are allowed. A surface vehicle with a winch is permitted. The choice of materials is left to each group.

- The robot may be powered by any device, elastic bands, electric motors, or other mechanical contraptions. Part of the challenge is for the team to optimize robot selection and design.

- The robot must travel to the designated weight, connect to the hook on the weight, and bring the weight back to the surface at the pool edge.

- The robot must be tethered to the shore. A land controller, which may contain the power supply, navigation, and ballast equipment, can serve as the tether; however, the tether may not be used to pull or drag the robot or the marker string. Any power supply must be internal to the robot or the controller, i.e., not plug into a wall outlet, and no large external batteries are allowed.

- The robot will be placed at the edge of the pool and released by the group members. Each robot must function remotely, that is, it cannot be touched by a group member while operating other than from the control panel. If touched, pulled, or dragged, it must be recovered and restarted.

- Prior to the actual challenge, each robot must complete a successful run in the test tank. Each group should document the robotic performance: e.g., ballast and steering control, power, ability to lift the weight.

- On the challenge day, the robot will have ten minutes make a successful recovery.

Developing a Test Program

Engineering design and development requires validation and testing. Your notebook must clearly identify a development schedule, test dates, and test results. Modification to the design after each test must be documented. The final runs must be clearly recorded and dated in the notebook and the final results reported on the summary page.

For example, consider buoyancy design. This can be done with a flotation device or by using downward propulsion. Several trials may be required to select and test a suitable design.

4.8.4   SAFETY

The objective of the challenge is to foster engineering creativity and cooperation. The design group is responsible for ensuring safety of participants and spectators during the challenge. Groups using any feature deemed dangerous by the judges may be asked at any time to prepare a safety plan or suitably modify the design before continuing in the challenge. Offending designs may be disqualified at the discretion of the faculty member or peer assistants. Use of pyrotechnic or similar devices is strictly prohibited. Any questions regarding safety may be directed to your instructor.

The pool deck is wet and slippery. No shoes are allowed to be worn on the deck. Because the deck is wet, plug in power is prohibited.

4.8.5   CHALLENGE ORGANIZATION

Early Organization:

- Identify the pool, and schedule the design challenge day.

- Arrange for life guards.

- Identify a practice area and setup.

Preparation for challenge day:

- Prepare summary score spreadsheet.

- Prepare data sheets and coordinate individual score sheets with the summary spreadsheet.

- Draw lots for the section challenge times.

- Prepare a press release if appropriate.

- Prepare a descriptive poster if the challenge is in a public area.

Challenge day: Peer assistant assignments

- Set up a registration table: log in each group, verify the power supply rules, assign a target number, and provide the individual data sheet: Typically, two assistants.

- Check design notebooks: Typically one or two assistants.

- Recovery targets: Typically, two assistants. The assistants confirm the target numbers are visible and replace them in the pool after recovery distance was traversed.

- Data recording: One or two assistants enter the data into the summary spreadsheet. One person may have to go back to the other activities to confirm data, so two people are preferred.

- Oversee site clean-up: all.

Following the challenge:

- Provide the summary data sheet to all instructors.

- Arrange any awards, e.g., pizza party.

4.8.6 CONCLUDING COMMENTS

There is nothing like water to compromise the best design, and this challenge amply demonstrates that fact. Three strategies emerged. Submarines trawled the bottom. Surface ships trawled for the weights then tried to bring them to the surface, and, finally, surface ships with winches attempted to grapple the weights, pull them to the surface, and return them to the edge of the pool. Each weight had a string and a ping pong ball hot glued to the string. Ten weights were spread around the pool and each section had numbers 1–10 assigned to the group for retrieval. The distance from the edge of the pool to the weight was similar for all groups. The pool has a current. That current was sufficient to keep underpowered boats from reaching the weights. Our fall guest speaker was the Manager for Gulf well delivery for Shell Oil, and this project was tied into the oil spill recovery efforts.

Figure 4.18: Redesign consultation time.

4.9 WIND TURBINES AND WIND POWER GENERATION

4.9.1 FACILITIES

This challenge requires a fairly large open space. Three test facilities were set up with large fans providing the wind. We used an auditorium for the challenge although a large conference room would work equally well. The fan was set up on a table and the distances marked on the table with masking tape. The Electrical Engineering Department fabricated a power meter to record the wind turbine output. This challenge is suitable for a public location.

4.9.2 CHALLENGE

This challenge requires each section to form its own company to manufacture and test a series of prototype wind turbines to generate electrical energy.

The national energy policy has a goal that 20 percent of all electric energy produced in the US should come from renewable sources by 2020. This has led to considerable development of wind farms across the country. Wyoming is one of the premier wind locations in the US and a substantial

Figure 4.19: Wind turbine testing.

amount of research into wind energy is conducted at UW. Just how efficient is wind energy? This design challenge is intended to let you examine the design issues associated with efficient wind generation.

Each group will design and construct a wind turbine to maximize the energy output and to determine the efficiency of the design. To complete the challenge, each group must complete the following tasks:

1. Design and fabricate the wind turbine,

2. Set up the wind turbine 4 feet in front of the industrial fan,

3. Measure the wind speed immediately in front of the wind turbine (instrumentation will be provided),

4. Compute the theoretical energy input of wind,

5. Attach the power meter and measure the energy output of the turbine,

6. Repeat steps 3–5 with the wind turbine 6 and 10 feet in front of the fan,

7. Compute the efficiency of the wind turbine for each wind velocity and plot the resulting data,

8. Enter all data and test results in your design notebook, and

9. At the design challenge, demonstrate that your turbine can generate the energy output reported from your development program at 6 ft. from the face of the fan.

4.9.3 THE RULES

Each freshman Section will function as a team. A team consists of eight groups of two to three students. Wind turbines are to be designed, built, and tested to optimize the team's point score. In that regard, the team must decide what constitutes its best composition of turbines.

Scoring will compare the instantaneous energy output of all wind generators in the team. The highest average energy output receives the outstanding team award.

Individual groups present their ideas in class and receive input from the team.

- Each team must design and fabricate at least one vertical axis and one horizontal axis turbine. The remaining turbines are at the discretion of the team (Figure 4.20).

- Each group will receive an electric generator and a gear set. This generator must be used; the gears are optional. Gears may be traded within a team to optimize performance.

- Each group must maintain a design notebook (see Chapter 2).

- In addition to the parts supplied, no group shall spend more than $20.00 for supplies and equipment to manufacture its wind turbine. It is OK to use "free stuff." The definition of "free" is that it has no commercial value. In short, the instructor can elect to keep it or throw it in the trash following the competition. "Rented" or "borrowed" materials are not allowed.

- All wind turbines must be designed, fabricated, and tested prior to submittal.

- The turbine must be constructed from scratch, that is, no premade units can be used. The choice of materials is left to each group.

Wind Turbine Design

1. The cross-sectional area of the turbine blades (perpendicular to the wind direction) must be less than 324 square inches (18 in square or 10.2 inch radius).

2. The center of the wind turbine should be about 27 in. above the table—roughly at the center of the fan axis.

3. Bricks will be available to anchor the wind turbine base. The bricks should not extend more than 3 in. above the table.

4. You may construct a gearbox out of plastic sheet available in the shop.

5. The College of Engineering and Applied Science shop is available for fabrication of your turbine.

(a) Horizontal Axes   (b) Vertical and Horizontal Axes

Figure 4.20: Sample wind turbines.

Generator and Transmission

The generator is a 6 volt DC motor supplied in the parts kit. The optimum drive speed is 2,180 rpm, so a step up from the turbine blade speed may be needed. Each generator must have a 4 ft. long 2-wire conductor (available in the shop) attached to the generator for connection to the power meter.

Data for the generator is given in Table 4.1.

Table 4.1: Generator Technical Data: Nichibo DC Motor—FE-260-18130 Available from Jameco Inc.

Nichibo DC Motor – FE-260-18130 Available from Jameco Inc.			
Current @ Max. Efficiency (A)	0.08	Size (Dia)	1.259 x 0.767
Efficiency	50.6	Speed @ Max. Efficiency (RPM)	2180
Nominal Voltage (VDC)	6	Terminal Type	Solder
Shaft Diameter (inch)	0.078	Voltage Range (VDC)	1.5-12
Shaft Length (inch)	0.385	Torque @ Max. Efficiency (g-cm)	11.6

Each kit has a set of gears similar to those in Figure 4.2. The design may include assembling the gears into a gearbox with one axle to the turbine blade and an output axle to the generator.

4.9.4 DEVELOPING A TEST PROGRAM

Figure 4.21 shows a schematic of the turbine test setup. One test setup is available for testing and measurement prior to the challenge. A schedule of lab availability times for testing will be posted.

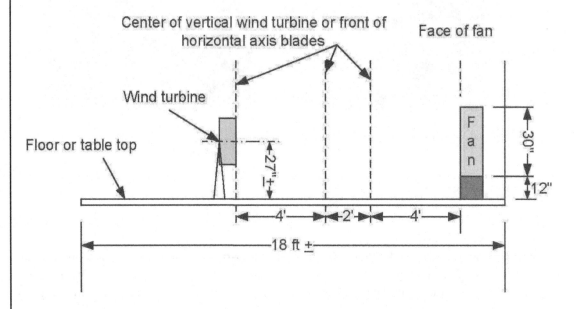

Figure 4.21: Schematic test setup.

Each group must establish a test program for their turbine. The wind turbines are to be set and tested at 4, 6, and 10 feet from the fan. The vertical axis turbine is set with the axis at the 4, 6, and 10 ft. marks. The horizontal axis turbine is set with the face of the blades positioned at 4, 6, and 10 feet from the face of the fan.

Turn on the fan and measure the wind speed and compute the theoretical wind energy. The fan will be positioned on a 12 in. high pedestal to make the centerline of the fan axis at an elevation of 27 inches and to reduce ground effects.

A wind velocity meter and the power meter will be available for testing.

4.9.5 CALCULATIONS

The average mass of air at 7,000 ft. (2130 m) is 0.0620 lbm/ft^3 (0.9920 kg/m^3). The total kinetic energy from the wind is $1/2\, mv^2$, where the mass of the air is that swept by the turbine. The output energy of the wind turbine is measured with a power meter. Electrical energy, P, is computed by:

$$P = VI$$

Where P is energy in joules, V is voltage and I is current.

Efficiency is computed: $\eta = 100\% \frac{Energy-out}{Energy-in}$

where the energy-in is the theoretical wind energy and the energy-out is from the power meter data. Be sure your units for the kinetic energy and electrical energy are compatible when you complete the efficiency calculation. Compute the energy-in and -out for the three turbine locations and then plot the efficiency versus wind velocity.

4.9.6   CHALLENGE DAY SETUP

Prior to the start of the official challenge, groups must complete "calibration tests." design notebooks must be submitted on the challenge day.

Each group will have five minutes to set up the turbine and measure the output. On the challenge day, all tests will be run at 6 ft. from the face of the fan. At the judges' discretion, any turbine not ready after the five minute countdown will forfeit and an energy output of zero will be recorded. The group may be allowed to retest after all other teams have completed their test.

At the start each test, the judge will indicate start and the clock will begin. The test will last until the energy is recorded or time expires. Groups may maintain contact with turbines prior to "go" but may not touch the turbine during the test interval. Turbine blades must be stationary when the test begins and the fan is turned on. A group member may touch the turbine to adjust parts but the energy record must be completed without manual assistance or the test will be disqualified.

Only the called group may enter the stage challenge area and the turbine repair area. Only team members may request verification of opposing turbines for compliance with contest rules and design limits. Spectators are not permitted to make such requests.

Any challenge must be made to the judges prior to the awarding of points for a particular test. If a turbine is challenged and found to violate context requirements, it will be disqualified and an energy output of zero recorded.

Scoring challenges for a particular round must be made by the end of that round of the challenge, and may only be made by team members. Resolution to point challenges will be made at the discretion of the judges.

No tools will be supplied on the challenge day. teams are expected to bring all necessary items to repair or modify turbines during the challenge, including spare parts.

After the official start of the challenge, only registered groups will be allowed in the challenge and work areas. Spectators are welcome to view the challenge from the seating area in the auditorium.

After each test and prior to leaving the test area, groups are responsible for verifying that point totals have been correctly recorded by the challenge judge. Judges will be instructed to oversee these checks. One group member will sign each score sheet.

4.9.7   SCORING

Scoring for the project will be based on completing tasks on time. Table 4.2 indicates the task and timing. The challenge requires the model to be completed, calibrated, and operational. The maximum energy output will be recorded for each group. The team score is the average of the maximum energy output from all groups in the team.

Table 4.2: Challenge scoring

Task	Date	Task Points	Total Points
Design Notebook progress	Design Day –Week 4	10	10
Have a model prototype that is described in the design notebook	Design Day –Week 4	10	20
Energy Generation Curve	Challenge Day	20	40
Functioning Model which Corresponds to Design Notebook	Challenge Day	10	50
Energy Prediction within 25%	Challenge Day	10	60

4.9.8   SAFETY

The objective of the challenge is to foster engineering creativity and cooperation. The design group is responsible for ensuring safety of participants and spectators during the challenge. Groups using any feature deemed dangerous by the judges may be asked at any time to prepare a safety plan or suitably modify the design before continuing in the challenge. Offending designs may be disqualified at the discretion of the faculty member or peer assistants. Use of pyrotechnic or similar devices is strictly prohibited. Any questions regarding safety may be directed to your instructor.

4.9.9   CHALLENGE ORGANIZATION

Early Organization:

- Identify a challenge facility track and schedule the design challenge day.

- Identify a practice area and setup.

Preparation for challenge day:

- Prepare summary score spreadsheet.

- Prepare data sheets and coordinate individual score sheets with the summary spreadsheet.

- Procure an handheld anemometer.

- Procure a power meter and make sure it is calibrated for the generator specified in the challenge.

- Draw lots for the section challenge times

- Prepare a press release if appropriate.

- Prepare a descriptive poster if the challenge is in a public area.

Challenge day: Peer assistant assignments:

- Set up tables, fans, and distance marks prior to students arriving.

- Set up a registration table: log in each group, grade notebooks, and provide the individual data sheet: Typically, two assistants.

- Power output: Typically, two assistants. The assistants confirm the power output of each group.

- Data recording: One or two assistants enter the data into the summary spreadsheet. One person may have to go back to the other activities to confirm data, so two people are preferred.

- Oversee site clean-up: all.

Following the challenge:

- Provide the summary data sheet to all instructors.

- Arrange any awards, e.g., pizza party.

4.9.10 CONCLUDING COMMENTS

The fans provided a reasonable wind velocity for the test; however, the power meter was somewhat overdesigned so we were always reading the low end of the scale. For the final challenge, we placed foamcore panels around the edge of the fan to better direct the wind toward the turbines. Virtually all the turbines measured some output. The gear box is essential to the success of this challenge. Scrap Plexiglas and Lexan sheet pieces, adhesive, and axle materials were available in the shop for "free."

CHAPTER 5

Interdisciplinary Design

5.1 OBJECTIVES

Goals of both ABET and the College of Engineering and Applied Science is to offer and evaluate comprehensive senior interdisciplinary design projects. Interdisciplinary in this instance means students from different departments are recruited to design a project as opposed to teams of students from within a single department. The following addresses how such projects are organized and executed.

5.2 ADMINISTRATIVE ISSUES

The first major administrative issue of a interdisciplinary project is addressing the individual departmental requirements for senior design. The University of Wyoming senior design requirements are summarized in Table 5.1. Initiation of an interdisciplinary team requires negotiation with each department to assure that the students receive proper credit, that workloads are commensurate with the credit hours, and that each department is satisfied that the project is commensurate with existing departmental protocols.

Table 5.1: Departmental Senior Design Requirements

Department	Fall Term Credit Hours	Spring Term Credit Hours
Civil Engineering	3 hours	
Civil Engineering-Transportation	2 hours	1 hour
Electrical Engineering	2 hours	2 hours
Energy Systems	3 hours	3 hours
Mechanical Engineering	3 hours	3 hours
Chemical Engineering	2 hours	4 hours

Further compounding the credit hour requirements are the departmental deliverable requirements. Civil and Chemical Engineering students prepare project plans and require oral presentations of their projects. Electrical and mechanical students must go through a product development process, fabricate their project, and prepare poster presentations.

Establishment of a separate section within each department for the interdisciplinary project resolved the departmental credit hours issue. The entire class meets as a group even though the students are all in "different" classes. While straightforward, the solution generated a second set of administrative issues. The university has a policy that undergraduate classes must have at least 10 students enrolled for the class to proceed. No department had that many students in one section. The college successfully petitioned an exemption on that basis that the entire project had sufficient student population.

5.3   PROJECTS

Six different interdisciplinary projects are presented. Two projects were tied to the NASA Zero Gravity research initiative, and the remaining four were local projects. The local projects required 1) design of an automated transit system for the campus, 2) develop more environmentally sustainable solutions for accessing gas fields in the state 3) converting the university energy plant from coal to wood, and 4) a non-engineering interdisciplinary course on medieval construction in conjunction with the History Department. The projects are summarized below and the projects are presented in the following chapter.

5.3.1   NASA ZERO GRAVITY PROJECTS

The NASA Zero Gravity program allows students to propose projects that would assist NASA in space endeavors. Students prepare proposals, submit them to NASA, and if accepted, construct the prototypes. Students then take their projects to the NASA center in Houston, Texas. After review and training by the NASA staff, the projects are loaded into the NASA KC-135A, and the students conduct their experiments in a weightless environment.

The first Zero Gravity project challenged students to develop truss elements and connections for the construction of three-dimensional truss structures in space. The project consisted of designing carbon fiber truss elements and plastic connectors. The second Zero Gravity project involved developing a zero gravity exercise machine for use on the international space station.

5.3.2   AUTOMATED TRANSIT SYSTEM FOR CAMPUS

The students were asked to design an automated transit system for campus that linked the remote parking lots with the main campus. Completion of the project involved traffic studies for both parking and walking distances from stations, development of the vehicle concepts, vehicle design, guideway design, power supply, and operations. The project was conducted over a two-year timeframe.

5.3.3   DISAPPEARING ROADS AND GAS EXTRACTION

Jonah Field in western Wyoming was one of the first major gas fields developed by fracking processes. Aerial views show a spider web pattern of roads in a region that is highly sensitive to pronghorn migration and sage grouse habitat. The design project required the students to interact with energy companies and the Bureau of Land Management to develop alternative designs to reduce the site impact.

5.3.4 UNIVERSITY ENERGY PLANT CONVERSION

The University of Wyoming Energy Plant provides steam heat to the campus. Originally constructed to burn anything from garbage to coal, it has functioned as a coal plant since its inception. Recent years have seen an extraordinary amount of beetle kill in the nearby national forest. The student project evaluated if the plant could be converted to wood chips fuel source, established that sufficient wood was available to support the conversion, designed handling and transportation of the wood, developed plans for modification of the energy plant, and participated in the first pilot burn.

5.3.5 MEDIEVAL CONSTRUCTION

This was a joint project with the History Department. Dr. Kristine Utterback of the History Department presented the life and times of Medieval Europe. The engineering aspect of this project focused on the design and construction of the first Gothic Cathedral in St. Denis, France.

5.4 STUDENT RECRUITMENT

The general concept of each design project is developed to the summary level indicated above. The project is open to all seniors needing to fulfill a senior design project. The projects are run for an academic year beginning in the fall term. To suit the departmental requirements, Civil Engineering students completed their work in the fall and a new group of civil engineering students joined in the spring. Other students enrolled for the full year.

Critical to recruiting students is the selection of the project. Student engagement was highest when the project either directly addressed an issue the students recognized or had a clear environmental benefit. One year several national competitions were suggested in addition to a local project. The local project was overwhelmingly the student choice. The professor in charge of an interdisciplinary project must be able and willing to adapt to the projects that will attract students.

Consideration was given to using an *Engineers without Borders* project. *Engineers without Borders* was not selected because it was not clear that all of the interdisciplinary design criteria could be satisfied. In addition, *Engineers without Borders* has projects underway and the schedules did not mesh.

Each spring, in the week prior to advising and two weeks prior to students signing up for fall courses, an announcement of the project is emailed to all seniors, and posters are placed in the college. A project open house for the project is held. A general overview of the project is presented at the open house and the students are invited to ask questions to explore their interest. It is only after the students sign up for fall courses that the composition of the class is known.

Supplementing the general sign up for the course, certain "specialists" are recruited directly. For example, one junior student in the chemical engineering department was mobility impaired and served on the university committee for campus mobility. After discussions with the department head, she was allowed to take this class early and had the responsibility for addressing all ADA requirements for the campus transit system design in addition to her technical responsibilities. Similarly, for the disappearing roads project, an Environmental and Natural Resources student was recruited to be the in-house environmental specialist. In both cases, these students added immeasurably to the quality of the project.

To further expand the class horizons, external classes are recruited to assist the design team. For example, on one project a senior class in Computer Graphics assisted in conducting focus group studies and preparing graphic imagery for the transit system alternatives.

5.5   PROJECT ORGANIZATION

Two key elements determined how the interdisciplinary design class was organized. First, the composition of the class had to be established. Because the class was open enrollment, the number of civil, mechanical, electrical, and chemical engineers is not known until the end of the spring term. Once the composition of the class is known, the actual project is modified to fit the class composition. Second, during the first week of class in the fall, each student is interviewed to establish the goals and aspirations the student has for the course. Prior to the interview, each student fills out a data card with their name, major, specialized skills like specific computer programming experience, GPA, and a statement on what they expect to gain from the course. From these interviews, the project is further adjusted to meet the skills and majors of the available students. A project manager and project engineers for each of the key components are selected.

The class is structured similar to a design office. The professor becomes the "principal in charge," and the day-to-day work responsibilities are given to the students. The project manager is immediately charged with establishing a schedule and works with the professor to assign students to the individual tasks. During class periods, the professor meets with the project manager, each lead engineer, and individual students to review progress. Each team member is expected to provide written or oral weekly updates on progress, difficulties, and resources needed to complete the task. The transit system description in Chapter 6 contains a representative organization chart. All senior design projects require written final reports and public presentations.

The professor's job is to assure that the project stays on schedule, redirect the project if it is going off track or the students propose solutions well outside their capabilities, and to arrange for supporting materials. The supporting materials include anything needed to design and fabricate the project, meetings with sponsors, field trips, or access to key personnel on or off campus. These resources vary with the project. Classes have guest presenters ranging from U.S. senators, to the university president, to maintenance staff. Every project involved at least one field trip.

Each project required some level of funding. The funding came from corporate support, the dean's office, or the H. T. Person Endowment. Individual funding is discussed in the project descriptions. Securing funding support is the responsibility of the professor.

5.6 ASSESSMENT

Multiple assessments of the projects were conducted. The NASA Zero Gravity projects were assessed by whether NASA accepted the project, and the NASA critique of their final report. Each project required the student teams to make a public presentation of their project. Members of the university, industry, professional engineers, and the press are invited. Everyone in the audience is given an assessment sheet and asked to critique the presentation. Professionals associated with the development of the project or the field trips are invited to the presentation and are given copies of the written report in advance of the oral presentation and asked to critique the presentation. A sample critique summary is provided in Appendix IV.

The work demands of these classes are one of the biggest challenges of their academic careers. The interdisciplinary designs have been well received by the students. They became the best advocates for the next year's class and many used their final design report as an example of their work in job interviews.

CHAPTER 6

Interdisciplinary Projects

6.1   INTERDISCIPLINARY DESIGN PROJECTS

Each of the following interdisciplinary projects follows a parallel format. The objective of the design project is presented, the student composition is given, the course organization is presented, the student results summarized, and an assessment in the form of closing comments are provided.

6.2   NASA ZERO GRAVITY I: CONSTRUCTION IN SPACE

6.2.1   OBJECTIVE

This Zero Gravity project challenges the students to develop techniques and connections for the construction of a truss structure for space applications. The project consists of designing carbon fiber truss elements and plastic connectors, practice assembling the truss, developing safety plans, and executing the design in the NASA KC-135 aircraft.

6.2.2   CLASS COMPOSITION

This class consisted of seven undergraduate students ranging from sophomores to seniors. Two were civil engineers and four were mechanical engineers and one was a senior journalism major. The two engineering seniors promoted the project as an independent study program to meet their senior design requirements. Dr. David Walrath, of the Mechanical Engineering department, provided technical assistance and coordination with the ME department requirements.

6.2.3   CLASS ORGANIZATION

The class was structured in a discussion format and given a single course number. The original concept of space construction came from the professor. The first quarter of the semester explored the range of possible solutions. The project was organized into tasks commensurate with the NASA project requirements. These included design of the experiment, detailed component design and fabrication, safety plans, and storage and material retention devices for on board the aircraft. Connection of the truss elements at the nodes emerged as the critical element of the project.

6.2.4   STUDENT WORK

The student design consisted of hollow carbon fiber truss elements and polyethylene connector units (Figure 6.1). The truss was to be assembled in the zero gravity environment onto base elements

Figure 6.1: Snap together truss design.

fastened to the top of a container box. The container box was fastened to the aircraft floor in accordance with NASA guidelines.

For comparative evaluation, two different truss concepts were developed. The first truss concept consisted of snap together elements. The second concept had a cam mechanism on the connector so that the node could be opened, a ball from the truss end inserted, and the cam twisted shut to complete the connection.

The zero gravity flight consists of a series of parabolic curves. Weightlessness occurs while the plane is in the parabolic arc. The gravity free time to work is on the order of 20 seconds. In order to prepare for the flight, the students practiced the truss assembly underwater in the campus swimming pool (Figure 6.2). This trial program worked out especially well because the amount of time available for holding their breath and working underwater was again approximately 20 seconds. The underwater fabrication occurred in a semi-weightless environment. The truss elements were close to neutral buoyancy and the students had no firm surface to grip.

The project was accepted by NASA and scheduled to fly in the spring semester. Each truss element and connection element had a Velcro strap attached to it and a corresponding Velcro tie

Figure 6.2: Underwater trial fabrication.

down on the container box. The Velcro prevented the pieces from floating free during the zero gravity portion of the flight. Both truss concepts were capable of being constructed within the time constraints of the flight. The final report to NASA concluded that the cam device was superior to the snap together design. While the snap together design initially could be constructed faster, the cam system was much easier to disassemble or reconfigure.

All components were designed and fabricated by the students. The aluminum storage box and the foundation of the truss were designed by the students and were common to both truss systems. The design included the safety padding along the edge of the truss container box (Figure 6.3).

6.2.5   ASSESSMENT

The project required two semesters due to the NASA review and acceptance process. Students presented their results to both NASA and to their respective student professional societies on campus. There was sufficient interest in the project that a second project was undertaken the following year. NASA was critical of the student final report in not being detailed sufficiently for deployment. In addition to the construction write-up, NASA indicated they wanted a full weight and strength analysis.

6.3   NASA ZERO GRAVITY II: EXERCISE MACHINE

6.3.1   OBJECTIVE

In zero gravity, astronauts need self-contained, self-reacting, light weight exercise equipment for transport to space and it needs to be compact enough for easy storage on the space station. The NASA flight must demonstrate the suitability of the equipment in zero gravity. Students designed a "Bowflex ®" based exercise platform that allowed upper and lower body exercises. The final device

Figure 6.3: Snap together truss construction in zero gravity. (Photo courtesy of NASA)

was designed to be collapsible and fit in a minimum volume for both transport to and storage on board the space station. The project organization and executions were similar to the truss construction project.

6.3.2   CLASS COMPOSITION

The class consisted of ten students, four civil engineers and six mechanical engineers. Four of the team were female students. Dr. David Walrath assisted with the mechanical engineering components.

6.3.3   CLASS ORGANIZATION

The class was structured in a discussion format and given a single course number. The original concept of space construction came from the professor. The class worked in a colloquium setting and was responsible for fabrication of all components and preparation of reports.

6.3.4   STUDENT RESULTS

The students examined a range of exercise equipment and eventually selected on modifying Bowflex® system components to provide a force resistance system. The Bowflex® rods were lightweight, and could be procured in varying stiffness. They could be configured to provide upper and lower body strength exercises (Figure 6.4). Elastic band solutions provided similar exercise options, but the rigidity of the student design allowed easier mounting and dismounting exercise positions.

6.3.5   ASSESSMENT

The project ended up being only partially successful. While the student design functioned as planned, one of the project sponsors did not want to continue due to potential liabilities of elements breaking in space. Breakage was never an issue in the trials, but the inability to potentially replace a part remained a concern to the team. Complementing the student effort, the college supported the travel expenses for a reporter from the local TV station to accompany the students to Houston. This resulted in a five-day TV documentary featuring the students, their project, and their flight.

6.4   DESIGN OF AN AUTOMATED TRANSIT SYSTEM

6.4.1   OBJECTIVE

The main campus of the University of Wyoming is growing, and parking is being relocated to the campus perimeter. An interdisciplinary senior design class was recruited to design an automated transit system for the campus. This was a two-year project. The first year involved planning and preliminary design of the system. The second year involved constructing a prototype model transit system and alternative guideway designs.

(a) Demonstrating leg strength   (b) Changing exercise setup

Figure 6.4: Flight team on board the NASA KC-135A test flight (Photos courtesy of NASA).

6.4.2   CLASS COMPOSITION

The class composition consisted of five to eight civil engineers, two chemical engineers, one electrical engineer and four mechanical engineers. One of the mechanical engineering students was a dual major in ME and EE. The number of students varied from semester to semester due to the civil engineering departmental requirements.

6.4.3   CLASS ORGANIZATION

This was the first fully interdisciplinary senior design project. The course was divided into five components. The first component required about half a semester and dealt with project planning. The second component refined initial concepts and selected the overall transit system. This component included field trips to assist in understanding the magnitude and complexity of the undertaking. The third component began in the second semester and included detailed design of the system elements including vehicle, guideway, geometric layout, stations, maintenance facility, and control center. This activity additionally incorporated graphic art consultants. The last two components occurred in the second year and included design of an alternative guideway and fabrication of an operational prototype.

The class was organized similar to a design office. Figure 6.5 provides the second semester organization chart for the class including the various project assignments. The graphics design portion of the project was provided by the senior *Computer Graphics II* class.

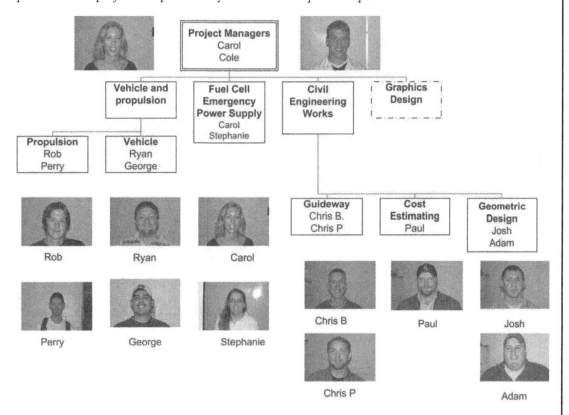

Figure 6.5: Class organization chart.

6.4.4 STUDENT RESULTS

First semester established the design parameters for the project. Five tasks were completed. First, the students reviewed material on automated transit systems found in the literature and lectures prepared by the professor. Second, they conducted traffic studies to determine the demand to be placed on the system, stations, and vehicles. Third, they examined the Americans with Disability Act requirements. Fourth, they developed the technical design guidelines. Fifth, they developed the preliminary design concept. Augmenting the literature review was a field trip to Denver, Colorado, where the team visited Rocky Mountain Prestress, Six Flags–Elitch Gardens, and the Denver International Airport Transit system. Rocky Mountain Prestress provided the team with insight to understand how construction could be prefabricated to minimize on site construction time and disruption. Six Flags–Elitch Gardens engineers discussed switching, safety, and operation of rides with small vehicles (Figure 6.6).

Figure 6.6: Discussing switches at Six Flags–Elitch Gardens and conducting bus traffic studies.

The Denver International Airport automated transit system requires very high reliability and has a sophisticated maintenance area. Students were introduced to transit operation reliability concepts.

The second task developed the overall load criteria and transit layout. This study included the size of the transit vehicles, frequency they would run, and the route they would take. To complete this task the students conducted extensive surveys of the shuttle bus. The university runs buses on 10–15 minute headways from about 7:30 in the morning until 8 in the evening. The student findings were enlightening. First, the only times the buses were heavily used were in the 20-minute period prior to 8 AM and 9 AM morning classes. Many times during the day the buses ran empty. Supplementing the traffic study, the students met with the president of the university to review the long range planning and capital construction plans. In addition to determining where the campus traffic would likely locate, they also examined athletic events on campus to evaluate if the transit system could assist access to football and basketball games. The study determined that the transit system would be elevated to eliminate conflict with street traffic.

As part of the traffic study, the students conducted an assessment of how far patrons would walk. From a series of transit studies in Canadian cities, they determined that transit users would walk about 1,000 feet before considering alternative mobility options (Figure 6.7). The present and future station locations provide access to 100 percent of the present and future campus.

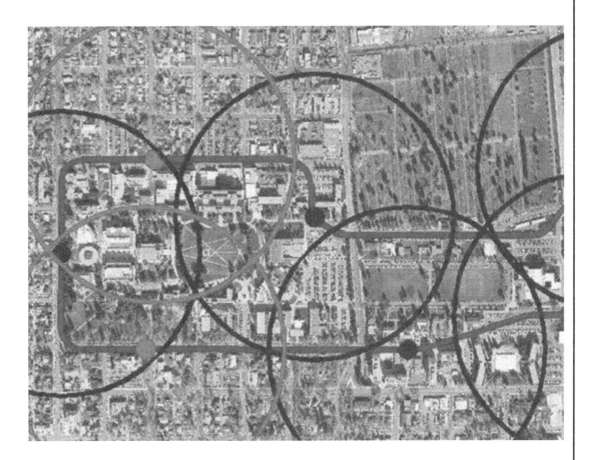

Figure 6.7: Aerial plan showing station locations and walking distances.

The third task determined how the Americans with Disabilities Act (ADA) would impact their design. In addition to reviewing the ADA requirements, the students laid out a mockup of the interior of the vehicle, then used wheelchairs to enter, exit, and position themselves in the vehicle. This study led to four major conclusions. First, all stations would require elevator service. Second, the vehicle must be able to accommodate one and preferably two wheelchairs. Third, wheelchair access must not require supplemental assistance or restraints. Fourth, the team must develop appropriate emergency egress solutions.

The fourth task established the design guidelines for the system. These were divided into two parts. The first part examined ASCE-7 *Loads on Structures*[1] for external loads on the structure and stations. ASCE-7 does not provide wind loads for guideways or vehicles, so the students had to extrapolate the specification data to suit their conditions. The second part addressed the vehicle requirements and consisted of two components. The first component was the overall frequency of vehicles, travel times, and interface requirements with the guideway to assure fatigue performance and ride comfort. The latter item leads to maximum horizontal accelerations and corresponding minimum curve radii. The second component was the size of the vehicle and its orientation on the guideway.

The conclusion resulting from the first semester was that the vehicle would be suspended under the guideway to mitigate the effects of weather in the Laramie area. An overhead support system assures that the bogie supporting the vehicle is within an enclosed area and out of the weather. While Laramie is semi-arid, the winter snowstorms and associated winds were a concern that an exposed guideway surface could become iced and result in the vehicle being stuck on inclined areas. To further assure all weather operations, a linear induction motor drive was selected. A linear induction motor not only provides the power climbing hills but also works to control downhill speed without having to resort to a mechanical braking system. The mechanical braking system for the vehicle served as a backup. Safety and emergency egress solutions were developed and included in the recommendations.

The traffic analysis suggested that vehicles with a capacity of six seats were adequate for the majority of the travel. The floor space was then designed to accommodate two wheelchairs. This provided a vehicle with a total capacity of approximately 20 students if wheelchairs were not present. That would be six students sitting and 14 standing. While this would be a relatively tight configuration, it was satisfactory to carry the peak load occurring just before the 8 AM and 9 AM classes. The students concluded that a small vehicle operating at a four-minute headway would be optimal for the peak hours. Vehicles would automatically be removed from the system and headways increased to 5- to 10-minute headways off peak. The students further decided that the transit system should offer two-way operation. That is, one side of the track would carry vehicles in a clockwise direction while the opposite side would carry them in a counterclockwise direction. The guideway would split at a station so the station would be between the two tracks and therefore only require one set of stairs and elevators. The two-directional operation assured that the minimum transit time would result and provide redundant operation. A rider could go counter flow to get to a station immediately across campus instead of having to ride the entire route to get to the same location.

During the second semester, the engineers "hired" the ART 4110 *Computer Graphics II* design class to assist in developing the system graphics. Three teams from this group presented graphic concepts to the engineers. *Stagecoach* emerged as the system theme. The graphics classes presented marketing concepts for advertising on the side of the vehicles to assist in defraying operational costs.

To publically assess the overall concept, the two classes conducted a focus group study. The study was conducted in the lobby of the student union (Figure 6.8). Students visiting the booth were requested to vote on a name, final graphic themes, and provide opinions on travel times and station locations.

Figure 6.8: Focus group booth and one schematic of a vehicle.

The guideway design was a steel truss spanning between precast concrete columns. The columns were designed to have a sandstone finish to match the buildings on campus. The final guideway design layout was selected to minimize the number of parking spaces taken and trees impacted. The suspended vehicle system had a secondary benefit that should one of the large cottonwood trees surrounding the campus lose a branch, that branch would hit the guideway but would not cross the guideway in a manner to disrupt or dislodge a vehicle.

The stations were designed in precast concrete and galvanized steel. The architectural finish was selected to match the buildings on campus and to be in accordance with the university trustees' guidance for overall campus architecture. The stations were intended to be modular to facilitate ease of construction. Where possible, it was also anticipated that the stations might be integrated into any new building construction to provide even more efficient access to campus facilities.

The mechanical engineering component of the project was satisfied by using the rapid prototype modeling equipment at the university. Students designed the overall cab for the vehicle and the bogie system. Each of the components was "printed" on the rapid prototype machine. They were manufactured to approximately 1/10 scale and were available for inspection during the public presentations.

Chemical engineering students were charged with developing a fuel cell component for each car to allow a vehicle to return to a station in the event of a power outage. Following their initial research, the students concluded that such power supplies were available commercially. They undertook a study for an alternative power supply for the project. The students developed a concept for using a solid oxide fuel cell currently under development by Siemens in Germany. The fuel cell operated on natural gas at a temperature of approximately 900°F. The fuel-cell generated sufficient electric energy to completely operate the transit system and have a 20 to 30% excess capacity to provide base load and off peak transit power to the university. The student analysis concluded that the heat from the fuel cell would be sufficient to replace the heat generated by the coal burning furnaces at the university energy plant and thereby reduce the university carbon footprint.

The projected construction schedule for the project was 700 working days from bidding until final construction. The project budget was estimated to be $64.3 million and a 15% contingency for future design cost increases.

The design was presented at a public meeting and included invited judges. Adjudicators included the president of the university, the vice presidents of Research and Facilities, the dean of Engineering, five faculty members, and three professional engineers. In preparation for the presentation, one of the students suggested that an animated graphic of the systems would be impressive. The class provided aerial views of the campus and graphics of the transit system to Mr. Brendan Dolan, who in turn generated a complete campus model in the computer game *Roller Coaster Tycoon*. The model was used in the presentation and included aerial views of the system and a comprehensive passenger's view from inside the vehicle as it circumnavigated the campus.

The third semester broke the project into two parts. The first part occurred in the civil engineering course for design of prestressed concrete. That class used the design guidelines developed the previous year to develop design an alternative guideway in precast-prestressed concrete. In the execution of the precast concrete design, the students were introduced to significant geometric design constraints due to the centrifugal force on the vehicles and the corresponding torsional effects in an inverted C shaped structure.

Mechanical engineers undertook the task of constructing a prototype transit system. During the field trip to the Denver International Airport, Logplan LLC offered to provide the university with a small section of the original Denver baggage handling system. The students used the bogie and guide rails from the baggage handling system as the basis for the demonstration system. They modified the bogie to support a Plexiglas cabin complete with operational doors. The students modified the LIM rail and LIM motor to correspond to the overall design criteria. The final project demonstrated the operation of the automated vehicle system. The vehicle pulled into the first station and cycled the doors automatically. Optical sensors checked for obstructions and recycled the doors if an obstacle was present. The vehicle traversed to the next station, cycled the doors, then shut down (Figure 6.9).

6.4.5   ASSESSMENT COMMENTS

The project was complex, difficult, and highly engaging for the students. It was assessed in two different environments. The first assessment was a public presentation of the project at the end of the first year. The presentation team included the engineering students and the *Computer Graphics* students. Evaluation sheets were given to everyone in the audience and specific sheets were given to

Figure 6.9: Students programming the model transit system.

individuals asked to adjudicate the project. Each critique sheet asked the reviewers to evaluate the project on the basis of the written report, the technical merits, and the oral presentation. On a scale of 1 to 4 with 4 being outstanding, the technical review team score was 3.6. Non-engineers rated the team somewhat higher, 3.8, than the technical reviewers. A sample evaluation sheet is provided in Appendix IV.

A press announcement was compiled by the graphics class, and press releases were prepared. The project received coverage in most newspapers in the state. Copies of the final report were sent to the board of trustees and several state legislators.

UW TV conducted the second semester interview and presentation and the tapes were release within the state. One of the interesting facets of the mechanical engineers design was the fact that the students were able to make the Denver Airport baggage handling system work.

Figure 6.10: Final project graphics (Courtesy ART 4110).

The second level of review occurred when the final report was distributed to the various field trip sponsors. Several comments were received; however, a most interesting critique came from a firm that conducts planning of specialty transit systems. They had acquired a copy of the report from the Denver International Airport. Their comment was that the students had not followed all of the relevant specifications for transit design. At the same time, the firm understood that working from first principles not just following specifications was a class objective. They then requested the names and contact information for every member in the class as they wanted to hire as many as they could.

6.5 DISAPPEARING ROADS

6.5.1 OBJECTIVE

Jonah Field in western Wyoming was one of the first major gas fields developed by fracking. The tight sandstone led to wells being placed in close proximity to each other. Consequently, a spider web pattern of access roads developed. The class was charged with designing methods to reduce the surface impact of drilling. In the process of the course, the class also elected to enter the "Disappearing Roads" competition sponsored by Halliburton Corporation and run by Texas A&M University.

6.5.2   CLASS COMPOSITION

The class consisted of seven civil engineers, 12 mechanical engineers, and one Environmental and Natural Resources student. The second semester introduced a new group of civil engineers. Two civil engineers elected to take the second semester as an elective credit to complete the project design.

6.5.3   CLASS ORGANIZATION

The Disappearing Roads competition was presented to the class as a model but not a requirement for the project. The first two weeks discussed the environmental and engineering issues to be addressed. The class then traveled to Pinedale, WY, for a three-day field trip. The trip included stops at the Halliburton facility in Green River, WY, and a briefing of environmental and regulatory constraints by the Bureau of Land Management in Pinedale, WY. EnCana Corporation arranged a full day tour of Jonah Field including briefings on fracking operations, gas recovery, disposal of drilling materials, and overall operations.

Following the field trip the class elected to enter the Disappearing Roads competition. The class was interviewed by Mr. Richard Haut, of the Houston Area Research Consortium, and, based on their interview, were allowed to compete.

The second semester included a field trip to the Questar Productions, virtual drilling facility in Denver, CO. The tour included a 3D visualization of the Pinedale Anticline Production Area and the difficulties of hitting the small gas formations.

Near the beginning of the class, a request came from U.S. Senator John Barrasso's office for a student panel discussion of energy policy based on the book *Beyond Oil*.[1] Six students from the class participated with one student serving as moderator. At the conclusion of the course the group that was representing the university at the Disappearing Roads competition presented their findings to Senator Barrasso. The senator met with the students for well over an hour and quizzed them closely on their work. This briefing was exceptionally helpful in preparing the students for their Disappearing Roads presentation.

6.5.4   STUDENT RESULTS

The following student results focus on the environmental considerations leading to development of a mat road system. The work included research, development, testing of concepts, and concluded with the class participating in the Disappearing Road competition.

The Pinedale Anticline Production Area (PAPA) and Jonah Field are in West Central Wyoming west of the town Pinedale. Though these two fields share many of the same characteristics, there are a few key differences. First and foremost is the size of the drilling field. PAPA encompasses 198,000 acres. This is over eight-and-a-half times the size of Jonah Field. The PAPA is a long narrow swath of land that stretches from Pinedale to 70 miles north of Rock Springs. The

Figure 6.11: Class briefing U.S. Senator John Barrasso.

terrain at the PAPA is generally not as level as the terrain at Jonah Field. The PAPA does have a very similar dry climate to Jonah Field. The fields contain over 3 billion cubic feet of natural gas.

Common environmental concerns in PAPA and Jonah Fields include: impacts on sage grouse, pronghorn antelope, big game animals, top soil disturbance, air pollution, preserving view sheds, soil chemical composition, and addressing archaeological issues.

While the two sites share many of the same environmental concerns, the PAPA has far more habitat concerns. The PAPA is a vast area, and it is broken into nine separate management areas that are based on land ownership and environmental concerns. Each management area has a set number of wells that can be developed and its own set of environmental concerns. These concerns range from preserving historic wagon trails to protecting big game winter habitat. Because of these additional concerns different strategies will have to be applied to the two natural gas fields.

Figure 6.12: Aerial view of Jonah Field (Photo copyright Jeff Vanuga, used with permission).

The Pinedale Anticline Project Area and Jonah Field have many shared geological traits. In both fields the natural gas being recovered is contained in over-pressurized pockets in the Lance Formation. These pockets in the Lance Formation are described as a bowl of potato chips. Each chip contains the gas bearing formation. These pockets require hydraulic fracturing in order to recover the gas. Hydraulic fracturing sites require a larger and heavier footprint than a conventional natural gas well site.

A key to reducing reclamation time is to reduce the disturbance of the topsoil. When the topsoil is torn up, and stored in piles for several years, the soil loses vital microbes and nutrients. The root structure of the sage brush is also heavily damaged in this process. Sagebrush can take 10 to 30 years to reestablish but if the root structure is preserved, the recovery can be as little as two years. Implementing strategies that would lessen the disturbance of the topsoil and sagebrush would lead to a reduced recovery time which would be beneficial both to the environment and to the energy companies as the regulations will only allow more drilling when an equal area is recovered.

Both Jonah Field and the PAPA are on land once inhabited by indigenous cultures and still maintain very important archeological sites scattered throughout the fields. Damage to these sites should be avoided, and strategies to account for these sites implemented.

Additional regulations that affect PAPA and Jonah Field include:

- In order to preserve air quality in limiting NOx emissions, all drill rigs shall use natural gas powered engines with low NOx emissions.

- Mat roads and pads cannot be in the same location for two years or more. If a mat is in place for two years at one location, it must be removed and cannot be placed in that location for another year.

- All compressing and condensate facilities must produce no more than 49 decibels of noise pollution.

- The owner of the land reserves the right to have any main road removed. If it is desired to have the main road removed, it shall be done by the energy company who must remove all foreign soil from the road system. A two track access road must remain for maintenance purposes throughout the service life of the well.

- If heavy equipment is required to access any site after development, a mat road must be deployed for access.

- All reclamation criteria will be according to the current BLM mandated reclamation criteria.

- In management areas with big game winter range there will be no development occurring from November 15 through April 30.

The Pinedale Anticline and Jonah Field have enough differences that they warrant two different strategies for drilling. This summary of the student work addresses the PAPA field. The recommendations for the Jonah Field are in their full report. The PAPA is slated to have about 2,000 wells drilled in the next 20 years and will be operating through the foreseeable future, whereas production at Jonah Field is on the decline, and suggested strategies may not be as applicable.

The PAPA has to account for additional time constraints due to winter range of big game and breeding season of the sage grouse population. Therefore, the solution in the PAPA is to limit the development footprint using temporary roads for field delineation. This research suggests a temporary mat road system would improve access and reduce recovery time.

The timeline for an individual well pad at the Pinedale Anticline is constrained as follows: no surface activity is allowed from November 15 to April 30 in management areas with big game winter range concerns. This leaves 198 days for development to take place. It is estimated that a well will be completed in approximately 72 days; at this pace five wells can be drilled in areas where operation can be year-round and three wells in areas where drilling operations must shut down in the winter. In order to complete one well pad with 32 wells it would take 11 years in areas with winter range concerns and seven years in other areas.

Due to these constraints, the timeline for a pad in an area with big game winter range concerns is as follows: on May 1, the mat road would be deployed. A temporary mobile modular frame would then be setup, with all equipment and material needed to complete three wells. Once all the equipment is in place, the mat road would be picked up, and a two-track road would serve as access for the workers. Once the final well is completed for the season, the mat road is redeployed and all the equipment that is required to leave the pad site would leave at that time. The mat road is then picked up and the site is vacated until the next year.

A main paved road would be designed for the spine of the PAPA field and should meet the following criteria: provide an adequate base, sub grade, and pavement type with a thickness capable of withstanding a repeated 80,000 lb truck load. A preliminary design indicates that the paved road would reduce dust, noise, and maintenance. It would be 6 in. thick consisting of a 3 3/4 in. nominal hot plant mix. The paved road would be at least 24 feet wide to allow for the larger turning radius of trucks and equipment. This scenario requires mat roads of up to one mile and would therefore need a complex mat road system. The mat roads would be at least 12 feet wide and as much as 24 feet where turns are required. Advantages of this scenario include not having to spray chemicals on a dirt/gravel road for dust suppression along with a smoother, faster, dust-free access to well and hydro fracking sites. The disadvantages of this system include a higher initial cost, and the requirement of a more complex mat system and maintenance access to the site. The gravel roads that spur from the paved road would be placed when a mat road is unable to connect a desired well pad site to the paved road because of safety or terrain reasons. Due to topsoil concerns, the maximum deployment for a mat road would be two years with a minimum of one year before redeployment in the same location.

A roll-out road concept is suggested for short access roads. The roll-out road incorporates hinged board segments linked with cables that can be rolled out into 50-foot road sections. These sections can be rolled out to construct a temporary road and then rolled up when finished. This concept will reduce the time required for setup and removal and enhance the ability to conform to uneven ground surfaces (Figure 6.13).

Figure 6.13: Roll-out road segment.

The key benefit of the roll-out road concept, compared to other alternatives, is the ease of placement and removal on site by incorporating a continuous roll rather than individual mat segments in a grid/matrix format. Parallel 10-foot-wide lanes allowing for a complete 20-foot-wide two-lane road to be rolled out. The four main components in the initial design included board selection, hinge design, segment connection design, and road dimensions.

During the class field trip to Jonah Field, representatives stated that the main problem with currently available mat designs was the longevity of the oak. The design team found a solution with Heartland Bio-composites, a Wyoming-based company that specializes in manufacturing natural fiber-reinforced/polymer-based lumber products (bio-composite). Some advantages in using bio-composite lumber are long-term durability, enhanced weather resistance, and their ability to be recycled. In following the "low impact/environmentally friendly" theme of the project, the design team felt that bio-composite lumber was an ideal solution. To enhance strength and minimize the number of segments required for the roll-out road, the team decided to base the initial design on a 2 x 8 in. board cross section.

With board selection completed, the next task was to maximize the ability of the boards to conform to uneven ground surfaces. The plan for the roll-out road called for individual board segments to be chained together in the longitudinal direction. Transverse hinges, centered in each board, hold the connection together in the direction of travel while still allowing individual segments to conform to changing terrain (Figure 6.14).

with hinge   without hinge

Figure 6.14: Lateral flexibility with and without hinge assembly.

Two conceptual hinge designs were developed for the roll-out road. The first hinge design incorporates a flexible elastomer/rubber hinge. The elastomer/rubber hinge is affixed to slotted board segments and held together with lag screws. The second hinge concept utilizes U-bolt fasteners and fabricated with ASTM A1018 steel endplates.

Before finalizing the roll-out road design, testing was performed on both an individual component basis and as a scaled prototype in the field. Testing can be broken down into "sandbox" board tests, hinge tests, and field tests.

Preliminary tests were run to assess the durability of the bio-composite boards by subjecting them to cyclic loading, which represents a continuous series of heavy duty vehicles driving over the road. To replicate field conditions, a "sandbox" was constructed and filled with a sand/soil mixture and placed under the hydraulic ram of a MTS machine. Oak and bio-composite test boards were continuously loaded with 4,500 lbs at a cyclic frequency of 1 Hz. Each material was tested for one hour (3,600 cycles) and the corresponding maximum deflections were recorded as 3.78 in. for bio-composite, and 3.28 in. for oak.

The second test administered in the laboratory was designed to test the tensile strength of the reinforced rubber and U-bolt hinge connections. Two types of reinforced rubber (Capralon® and Masticord®) provided by JVI Industries were used to assemble two separate hinges, both of which underwent tensile loading until failure. The U-bolt hinge was tested using the same procedure. The tensile loads at the point of failure for the Capralon®, Masticord®, and U-bolt hinge assemblies were recorded as 2,200 lbs, 1,450 lbs, and 4,800 lbs respectively. With a predicted maximum tensile load for the hinges of 780 lbs under field conditions, all three hinge designs performed reasonably well (Figure 6.15).

Figure 6.15: Hinge assembly testing.

The final testing application involved placing the prototype road section utilizing the Capralon® elastomer hinges in the field. The prototype was taken to Mountain Cement Company in Laramie, WY, where 80,000 lb twin side-dump trucks were continually driven over the road system on their way from the gravel/limestone quarry to the cement plant. The results of the field testing revealed some serious problems with the elastomer hinge design. After approximately four days and 153 truck passes in the field, two of the boards failed at the middle hinge connection. The failure was determined to be caused by stress concentrations in the notched cuts on the board ends, which encase the rubber hinge components. The design team feels that cold temperatures (nearing 0°F) also contributed to the brittle fracture that resulted in failure. Even with the elastomer/rubber hinge failure, the design received praise from the cement company as well as several drivers who thought the concept would be excellent if a better hinge connection could be implemented. The initial rubber hinge connection was abandoned and the U-bolt hinge design was chosen as a final design.

For the roll-out road system to function as intended, the following guidelines need to be followed. First, all large obstacles should be removed from the path of travel and the route brush-hogged. While the road should be able to conform to most terrain, not following the above recommendations will lead to premature failure of the road. For the placement and removal process of the road, a simple solution rolling and unrolling the road from around a forklift attachment was selected.

The process will allow the road to be rolled up and rolled out without ever having to drive directly on the terrain. When rolling out the road, the forklift will drive directly over the road section as it is unrolling. When rolling up the road, the forklift will be driven in reverse down the road section, allowing the forklift to remain on the roll-out road at all times. The weight of the forks, beam attachment, and beam is thought to be enough weight to compel the road to roll up. This process will become easier once an initial wrap is completed.

A replicate mat was designed similar to the wood mats currently in use but used a bio-composite material. The bio-composites were attractive for their potential lifetime over the oak mats used today and their ability to be recycled. The extra cost of using bio-composites is a concern. Therefore, these mats are designed to have a life cycle cost that is substantially less and have a longer lifetime than the wood mats that are currently in use today.

The complete layout of these bio-composite mats incorporated some ideas from the current wood mats with a few layout changes. The mats are 8 x 8 foot squares in order to be used both on drilling pad sites along with roads leading into these sites. The 8-foot width of the mat allows for a 24-foot road leading into the well sites, which is compatible with any size vehicle in Jonah Field.

6.5.5   TESTING

Various tests were completed for the bio-composite material and the prototype mat. The goal was to determine if the bio-composite material would be more conducive to the environmental constraints and more cost effective to the consumer. The tests were completed to assess how well the bio-composite material performed. The standard of comparison for the tests was oak, the material currently used in the field. The tests performed were for: friction, fatigue, abrasion, shear, deflections under loading, and a field test.

Concurrent with these lab tests, a field test was performed. Four quarter-scale mat prototypes (4 foot x 4 foot x 4 1/2 in.) were built and placed in a rock quarry road owned by Mountain Cement Company outside of Laramie, Wyoming, for field testing. Two mats were replicates of the oak mats currently in use and two were built with a 0°/45°/0° configuration to test for strength and durability for either configuration. These mats withstood an average of 2,400 tons per day for 18 days (Figure 6.15). After testing, both configurations came out looking exactly the same. There were no broken boards, the mats did not warp and they showed very little wear. The only problem that arose was when the mats were being removed, the interlocking boards ended up breaking because they were frozen to the ground and improper removal techniques were used. Because all mats were removed the same way, it is felt that the 0°/90°/0° composite prototypes should be pursued over the 0°/45°/0° configuration simply because they are easier and cheaper to manufacture.

Students recommended additional tests including an Izod test at extreme temperatures, a creep test on the prototype, the tests already performed with wider temperature variant conditions, longer field testing, properties testing to compare the theoretical properties to the actual properties, a screw withdraw test with various screws, and a shear test with various screws. They also recommended that

Figure 6.16: Field test.

a demonstration installation should be performed on Jonah Field to further test the bio-composite mats in the field.

The bio-composite mats are beneficial to use in the field over the oak mats. The bio-composite material is more environmentally friendly because it does not absorb significant amounts of moisture, it does not leach into the soil, and it can withstand varying temperatures. The bio-composite mats are also more cost effective because, even though there is a higher initial cost, the life of the bio-composite mat exceeds the oak mats and that the composite materials can be recycled by melting and reforming.

After the mat/rollup road is removed, there will be emergency field situations. Efficient ways of responding to these situations were addressed. Finding an efficient solution to an emergency situation requires weighing the effects of the emergency against the environmental effects.

In the situation where a worker is hurt, the first option is to drive out to the site of the accident. When there is a serious injury to a worker, there is always the option of driving off-road because at that point, the risk of injury or death greatly outweighs the environmental effects of driving off road. The second option is to use a helicopter from Flight for Life.

In fire or other such emergencies at one of the well sites, there are multiple options for handling the situation. If you need large equipment at the site, such as a crane, tracked vehicles can be available. These tracked vehicles are large enough to haul the necessary equipment to the site of the accident. An advantage of these tracked vehicles is that they will produce a minimal footprint, even while hauling other large equipment. When the fire needs to be contained very quickly, the only option is to drive fire equipment to the site, even if this means driving off-road.

6.5.6   ASSESSMENT

The above description addresses the design, fabrication, and testing of roll-out and mat road portion of the design. The students also designed at portable drilling pad structures to reduce the footprint. Presentations at the University of Wyoming were provocative with varying opinions being offered by the students, oil and gas industry, and BLM. The students defended their design well. The learning experience was enhanced by the differing opinions and positions of the reviews.

External assessment of this project included taking first place at the Disappearing Roads competition. The same year we received a request from Grand Teton national park on the project. The National Park incorporated wooden mats into their park improvement project using some of the recommendations in the student report. The mat road concept, using composite mats, has been incorporated into projects in Texas in part because of the Disappearing Roads report.

6.5.7   ACKNOWLEDGMENTS

The research team acknowledges the materials supplied by Heartland Biocomposites, Torrington, WY, and JVI Inc., Lincolnwood, IL. Technical support was provided by EnCana U.S.A. Inc. Pinedale, WY; Questar Production Inc., Denver, CO; Bureau of Land Management, Pinedale, WY; Mountain Cement, Laramie, WY; the University of Wyoming School of Energy Resources; the H.T. Person Endowment; and the College of Engineering and Applied Science shop and staff.

6.6   BEETLE KILL AND BIOMASS ENERGY

6.6.1   OBJECTIVES

The massive beetle kill of Lodgepole Pine in the nearby National Forest creates a hazard for fire, traffic, hikers, and power lines running through the forest. Roads and power line rights of way were to be cleared to 75-foot setback to prevent trees from falling on the roads or the lines. This design project assessed whether the timber that was being removed was sufficient to be an alternative energy supply for the University of Wyoming Central energy plant and the modifications to the Energy Plant to accommodate wood fuel.

6.6.2   CLASS COMPOSITION

The class consisted of nine students; six civil engineers, two energy system engineers, and one mechanical engineer. In the second semester only the energy system and mechanical engineers continued.

6.6.3   CLASS ORGANIZATION

The project was refined to match the available student skills. The University of Wyoming Central Energy Plant uses coal as the main fuel source for providing the energy needed to heat the University of Wyoming campus. The stoker-grate boiler employed at the Central Energy Plant has the ability to burn a variety fuels to produce the required energy needed. This project explores options for biomass use at the Central Energy Plant, reduction in emissions in relation to the University of Wyoming's emission goals, acquisition of beetle kill wood as a biomass energy source, the facilities required to store the wood, site design and layout for the energy plant modifications and additions, the environmental advantages of the project, a risk assessment, and the cost analysis of implementing such a project. The project focuses on the implementation of the cofiring solutions at the central energy plant including the following areas of interest: verification of the fuel mixture ratio entering the boiler, energy and economic forecasting for cofiring at the Central Energy Plant, and biomass cofiring combustion effects.

6.6.4   STUDENT RESULTS

This project was viewed as an economic opportunity for the University of Wyoming as well as an environmental opportunity. Currently the University of Wyoming campus is heated by a network pipes that deliver steam to individual buildings. The steam is produced at the Central Energy Plant where coal is burned in stoker boilers to produce the delivered steam. The University receives roughly 25,000 tons coal annually from the Grass Creek Mine of Thermopolis, Wyoming (Table 6.1). The University of Wyoming set a goal to reduce CO_2 emissions 15% by the year 2015 and of 25% by 2025. The University of Wyoming is striving to be carbon neutral by 2050.

The Medicine Bow and Routt National Forests are particularly susceptible to beetle infestation due to the morphology of the forest and mature trees that have had to withstand years of drought. These forests combined consist of approximately 2.7 million acres of which over 1 million acres, or nearly 40% of the forests, have been affected by the beetle epidemic.

Table 6.1: University of Wyoming Annual Coal Consumption at the Central Energy Plant from FY04-FY08

Fiscal Year	Amount of Coal Burned (tons)	CO_2 Emissions from On-Campus Coal (tons)
2004	24,097	57,926
2005	24,059	57,446
2006	24,297	58,221
2007	25,864	61,248
2008	24,510	58,165

Coal that is burned to heat the University of Wyoming campus contributes significantly to the overall carbon dioxide, CO_2, output of the University of Wyoming. The total carbon emissions of the university are 147,452 tons of CO_2, and on-campus coal use makes up 39% or 58,165 tons CO_2. These emissions have remained relatively constant from 1997-2009. By adding biomass as a fuel source at the energy plant there is a potential to move the university well on its way to meeting or even exceeding its emissions goals (Figure 6.17).

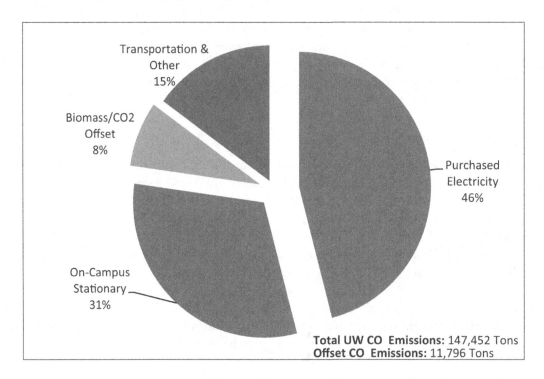

Figure 6.17: Estimated FY11 CO_2 Emissions Contributions by Source for the University of Wyoming with Cofiring 20% Biomass Replacement.

There are several environmental benefits to burning biomass in place of coal. In the case of emissions there are three clear benefits. They include: reduced sulfur dioxide emissions, reduced nitrogen oxides emissions, and reduced net carbon dioxide emissions. In this case, the sulfur and nitrogen oxides emission, reductions are minimal due to the low sulfur coal and the relatively small volume of coal being consumed. Therefore, the focus is on carbon dioxide emission reductions.

Burning biomass still produces CO_2 just as with any combustion reaction; in fact burning biomass produces almost the same amount of CO_2 as burning fossil fuels. Therefore, it is not intuitive that the biomass energy production is carbon neutral. To understand the emissions tradeoff for fossil fuels (coal) to biomass (wood) it is necessary to understand coal source. Coal is formed by plant matter from swamps that existed hundreds of millions of years ago that became buried; the plant remains became coal due to long-term exposure to heat and pressure. During its life, the plant absorbs CO_2, and CO_2 is stored in the plant matter as carbon. Therefore, when coal is burned it is releasing carbon dioxide captured by photosynthesis millions of years ago and is considered "new" carbon emissions. Biomass, on the other hand, releases CO_2 that was stored over the life of the plant, in our case several recent decades.

The trees killed by the pine beetle are no longer growing and thus no longer acting as a carbon sink through the process of photosynthesis. Over time these dead trees will naturally release the carbon that they have stored through the decay process or from a forest fire. If this carbon can be released during energy production and reduce the amount of coal that is burned, any CO_2 that is offset by the burning of biomass is considered a reduction.

A variety of solutions were considered for the use of biomass at the Central Energy Plant including: multiple offsite storage locations, four possible on-site storage areas, site modifications options, and boiler modifications for biomass-only fuel versus a biomass and coal cofiring solution. The team's final recommendations included the development of two offsite storage sites for long term biomass storage and processing at Centennial, WY, and Foxpark, WY; modifications to the north side of the current Central Energy Plant site to allow for biomass transportation and storage; and implementation of cofiring rather than boiler modification.

The decision to implement cofiring rather than converting a boiler to burn only biomass affected a variety of other aspects of the project and was therefore one of the first decisions required. A cofire solution evolved after reviewing the following decision matrix (Table 6.2). Boiler modification allows for the displacement of a larger volume of coal; however, this option has significantly more risk. By implementing cofiring the Central Energy Plant maintains flexibility to adjust for biomass fuel supply disruptions or price inflations, and a lower capital investment is required to begin the process. Based on GIS studies, two remote storage sites were selected in the Centennial and Fox Park locations. The sites total 18 acres, 14.5 acres dedicated to storage of wood and wood chips, 3.5 acres for contractor use to sort and load wood. The two-site solution provides flexibility for forest access and reduces transportation requirements as compared to a single site solution.

One of the concerns with introducing biomass to the current operations at the Central Energy Plant is the ability for the plant to store and handle the additional fuel while still maintaining the current coal storage and handling capacity. Maintaining the current coal capacity is important because it allows the facility to retain flexible operations and having enough coal reserve should a disruption in fuel supply occur during a period of high demand. The addition of a separate biomass storage and handling system allows the plant to maintain current coal and biomass capacity for greater flexibility to adjust fuel mixture ratios.

Given the need to mix fuels, several modifications to the Central Energy Plant site were proposed. The requirements for the biomass storage and handling system included: the ability to store up 100 tons of biomass, an unloading area for biomass that did not interfere with the current coal delivery system or other components of the site, and a tie-in with the current plant design. Option 4 on the north side of the site was selected as it provided the least amount of disruption to the existing operations while providing significant storage space (Figure 6.18 and 6.19).

The recommendation for the Central Energy Plant is to implement cofiring with the mountain pine beetle killed trees takes advantage of the environmental crisis and turns it into an opportunity.

Table 6.2: Risk Assessment Decision Matrix Comparing Boiler Modification to Cofiring at the University of Wyoming Central Energy Plant

Legend	
•	minimal risk
	moderate
√	risk
O	high risk

Risk	Option 1 Wood Boiler	Option 2 Cofiring
Wood availability or delay	O	•
Not enough wood supply	O	√
Must have carbon offset	•	•
Coal price highly fluctuates	O	•
Wood price fluctuates	O	√
Regular boiler maintenance	√	•
Boiler out of service	√	•
Storage area	√	•
Transport double handle	O	√
Variation in wood moisture content	•	•
Ash removal	√	•
Environmental disadvantages	•	•
Sustainability of project	√	•
Total=	5 - O	0 - O
	5 - √	3 - √
	3 - •	10 - •

Detailed design issues include: the ability to verify the fuel mixture ratio entering the boiler; the ability to predict the energy input into the boiler to account for any necessary control modifications, understanding how the addition of biomass will affect the burn chemistry within the boiler and how it relates to ash content and grate integrity as well as potential deposits on heat exchangers.

There are alternative ways to mix biomass with coal for the purpose of cofiring. At some plants coal and biomass fuels have been injected separately, while at smaller utilities they were mixed prior to injection. Premixing is done either on site or prior to delivery. At the Central Energy Plant, the pilot study determined that fuel mixing solution would store the woodchips to one of the three existing storage silos and then deposit them onto the conveyor flow with the coal. This wood/coal

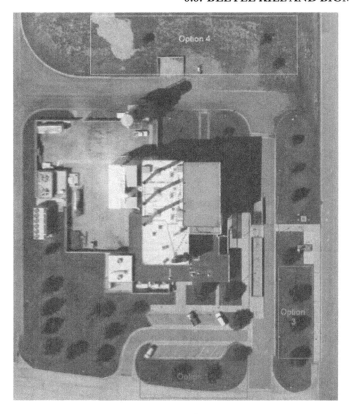

Figure 6.18: Aerial view of the central energy plant with the four proposed site modifications to accommodate for biomass storage and delivery.

mix will then be loaded into one of the three bunkers, which feeds directly into the boiler. The mixture ratio can be modulated by means of a guillotine valve at the base of the wood chip silo.

Moisture content is a limiting characteristic when implementing a biomass fuel source, for a particular biomass to be a viable fuel source. Three different samples of Lodgepole Pine wood chips were obtained from the Medicine Bow forest and tested for water content in the UW Civil Engineering soils lab. These samples included wood chips from a tree with no needles, a tree with red needles (dying), and one with green needles. The results from these tests are shown in Table 6.3.

These results are significant because the water content of beetle kill trees is sufficiently low that an expensive and cumbersome drying process would not be required. Low water content also contributes to the recoverable heating value and greater boiler efficiency. If green trees are harvested, they must be allowed to dry to reduce the moisture content below 20 percent.

The heat value of wood varies between species due to different chemical makeup. Most often the heating value is reported in units of energy per oven dry weight. When water is present, it

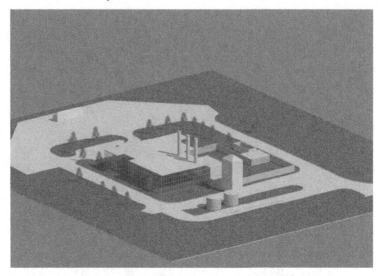

Figure 6.19: Rendering of biomass delivery and storage system.

Table 6.3: Water content of different samples of Lodgepole Pine

Sample	Water Content (by mass)
No Needles	8.83%
Red Needles	10.01%
Green Needles	57.91%

contributes a significant amount of weight to the sample, but not to its heat content. Also, when water is present in a sample that is to be combusted, there is a decrease in boiler efficiency because vaporizing water uses some of the heat that is liberated in the boiler.

It was found that the water content of wood is the most significant factor in determining the heat content of it. In fact, the per unit weight heat content of the wood decreases proportionally to the amount of water that is present. Although the heat content varies from species to species, 8,500 btu/oven dry pounds is an average for local wood fuels. The reported heat value for beetle kill trees was found by averaging the water content of the red needle and no needle trees and is approximately 7,600 btu/lb.

Completion of a successful test burn of biomass and coal cofiring at the Central Energy Plant required accurate determination of the mixture ratio of biomass and coal. An experimental program was developed to verify the mixture ratio of biomass and coal entering the boiler. The verification method needed to be simple and repeatable so that it could be performed by personnel at the Central Energy Plant at the time of the test burn and when the project goes forward. The procedure had to first be calibrated for the coal and wood chips used at the plant.

A "Mix Calculator" is a tool developed to process the basic information collected at the Central Energy Plant during the test burn and provide information about the mixture ratio (by volume and by mass), the estimated energy density of the fuel mixture, fuel cost estimates, estimated annual savings, and estimated annual carbon dioxide (CO_2) emission reductions. The embedded assumptions for the "Mix Calculator" can be found in Table 6.4.

Table 6.4: Assumptions embedded in the "Mix Calculator"

Assumption	Value	Units	Source
Bulk Density of coal	9.55	[lbs/gal]	Experimental data
Bulk Density of wood	2.35	[lbs/gal]	Experimental data
Energy density of coal	10500	[btu/lbs]	Central Energy Plant
Energy density of wood	7600	[btu/lbs]	Theoretical results verified by an independent lab
Cost Coal	56.00	[$/ton]	Central Energy Plant

In order to calibrate the "mix calculator" the following supplies are needed: a four-gallon sampling bucket, a scale capable of measuring weights up to 50 lbs, 300 lbs of coal from the Central Energy Plant, 35 lbs of dry wood chips from beetle kill trees, and two measuring buckets that can accurately measure samples as small as one half gallon. The procedure is:

1. Measure four gallons of coal, weigh the sample, and record the results.

2. Measure four gallons of wood, weigh the sample, and record the results.

3. Measure two gallons of coal and two gallons of wood; mix the components in the sampling bucket. Be sure that the mixture is homogeneous. Weight the sample and record the results.

4. Measure three gallons of coal and one gallon of wood; mix the components in the sampling bucket. Be sure that the mixture is homogeneous. Weight the sample and record the results.

5. Measure three and one half-gallons of coal and one half-gallon of wood; mix the components in the sampling bucket. Be sure that the mixture is homogeneous. Weight the sample and record the results.

6. Repeat steps 1-5 two more times to gather additional data points.

7. Calculate the bulk density of coal and wood for the different mix ratios.

This actual mix validation requires a four-gallon bucket and a scale capable of weighing up to 50 lbs. The following process should be repeated every three minutes upon changing the fuel mixture ratio until steady state is achieved.

1. Identify boiler which will be burning biomass-coal mixture and locate the sampling port.

2. Measure and record tare weight of the sampling bucket.

3. Gather a four-gallon sample from the sampling port into the sampling bucket.

4. Lightly shake the sample to allow for moderate settling and refill the sample to the four-gallon level if necessary.

5. Weigh the sample.

6. Subtract the tare weight of the sampling bucket.

7. Enter the result into the "Mix Calculator."

8. The output of the "Mix Calculator" will provide: volumetric mixture ratio, mass mixture ratio, the expected energy density of the fuel mixture entering the boiler as well as cost estimates for cofiring.

In order to use the "Mix Calculator" only three inputs are needed:

1. Cost of the biomass [$/ton] (Mountain Pine Beetle killed wood).

2. Current annual coal usage [tons/year].

3. Weight of fuel sample taken at the Central Energy Plant.

The above calibration and sampling method provides the Central Energy Plant a simple, quick, and effective method of measuring the mixing ratio and predicting the energy density of the fuel entering the boiler.

A review of the literature concerning cofiring coal with biomass revealed the technical issues related to the combustion process of these fuels. Table 6.5 is a summary of key differences of biomass compared with coal that should be considered when implementing a cofiring project.

Coal and biomass ash differ in terms of both their chemical and physical properties, as well relative amounts of ash produced during combustion. Further complicating matters is that interaction of these two fuels inside the boiler will have different effects on the predicted ash formation than if the two fuels were fired separately.

Typically, biomass has smaller ash content than coal with a different chemical composition. Coal ash is mainly composed of aluminum and silica, with clay and quartz, while biomass contains a high level of calcium and alkali metals such as sodium and potassium. It is the increased quantities of alkali metals that cause problems inside the boiler. These volatile compounds can act as fluxing agents, which, when combined with other mineral elements, cause them to melt. One example of this reaction is when potassium from the biomass combines with silicon from the coal and forms low melting silicates. These low melting compounds can bind fly ash materials onto the boiler tubes

Table 6.5: Comparison of Key Differences between Biomass and Coal

Characteristic	How Biomass Compares with Coal
Carbon Content	Lower carbon content than coal
Moisture	Higher moisture content than coal
Volatile Matter	Biomass contains a much higher percent of volatile matter and will de-volatilize independently of coal
Reactivity	Due to higher volatile content, biomass is more reactive and has a lower ignition temperature
Heating Value	Lower heating value
Ash Content	Lower ash content
Ash Composition	Higher volatile alkali metal content, especially in the form of potassium(K)

and fuel grates in a process known as slagging, the consequences of which include efficiency losses due to decreased heat transfer and air flow.

Alkali metals are much more prevalent in "rapidly growing" biomass such as wheat straw and switchgrass, while "old-growth" biomass, such as wood from pine trees, tends to have much lower alkali quantities. Accordingly, the use of Lodgepole Pine as a cofiring fuel will tend to cause less deposition problems than other biomass materials. Interestingly, the literature notes that one solution to the ash-related issues in biomass boilers is to cofire coal with the biomass. Some studies further suggest that the mineral elements in coal might have a "buffering effect" on the volatilization of alkali metals, which could further reduce the problem of fly ash deposition.

One other concern related to cofiring combustion is the chlorine (Cl) content of the biomass. This concern is due in part to the potential for Cl in the fly ash to corrode metal components of the boiler. Again, this is typically of greater concern for high Cl content herbaceous materials such as switchgrass and straw. Mitigation techniques for problems related to slagging and corrosion include increased soot blowing, the use of commercial chemical additives, ash deposition models, and attentive monitoring of the internal boiler conditions. A thorough chemical analysis should be performed on any type of biomass that will be combusted in order to understand the mineral composition of the resulting fly ash.

The final design included development of the offloading facilities, screw drives for wood chips, storage bins, and fuel mixing details. The calibration of the cofiring operation was completed. Following the conclusion of this project the Central Energy Plant continued pilot studies for including wood fuel and had dedicated one of the three coal storage bins to wood.

6.6.5   ASSESSMENT

The small class size limited the scope of the project. Energy systems engineers defined the quantity of wood needed annually for a 20-year cofiring conversion and designed the boiler modifications. Civil engineers conducted the timber availability and designed the staging areas and plant site modifications. Managers of the Central Energy Plant were judges on the final presentations.

The student design work was used by the Central Energy Plant, and the plant has continued to convert one boiler to a cofiring operation. The project served as the culminating senior design project for the first two students to graduate from the college's new Energy Systems Engineering program.

6.7   GOTHIC CATHEDRALS

6.7.1   OBJECTIVES

The course was jointly developed with Dr. Kristine Utterback of the History Department. Dr. Utterback focused on the life, times, and society of medieval Europe at the time of the first gothic cathedrals. The engineering effort focused on the state of knowledge at the same time of gothic cathedral construction. From an engineering perspective, this project presented engineering history to non-engineering students and introduced many engineering concepts used today.

6.7.2   CLASS COMPOSITION

The class was advertised on the university website, and a notice was sent to the local papers. The class consisted of 15 students, none engineering majors. Four students were non-traditional students. One criterion for the class was that only the math known in the Middle Ages was required. The lack of math requirements was a draw for non-technical students to any class dealing with engineering. The class composition included regular undergraduate students and a number of non-traditional students interested in the topic.

6.7.3   STUDENT RESULTS

A special summer course explored the development of medieval gothic cathedrals, which were built between about 1150 and 1400 in Europe. Bringing the vastly dissimilar areas of expertise, medieval studies and civil engineering, to bear on the subject gave the students very different perspectives on the life and times of the turn of the last millennium. The students explored the history of cathedrals as they developed in medieval society, examining the social, ecclesiastical, artistic, economic, and political elements. At the same time students planned and built a 1/8 scale model of a portion of a gothic cathedral, based on many of the same techniques medieval builders used. They began by conducting experiments on how arch and truss structures functioned. They continued by constructing their measurement tools, particularly the square and the level, using only a straight edge, a string, and a piece of chalk. The floor of the Kester Structural Research Laboratory became a "tracing room" as the students laid out the arches for the cathedral walls and vaulted ceiling.

The engineering instruction began by asking the students to build an arch. Wooden blocks were precut and the students had to assemble the arch. Many hands substituted for falsework. The frustration was high until one group "discovered" that an abutment was needed to support the horizontal thrust. Suddenly, construction progressed at a rapid pace (Figure 6.20).

(a) Attempted arch construction   (b) "Discovery" of abutements

Figure 6.20: Discovery of arch statics.

The next task fabricated tools. A line was struck on the floor of the structures lab. Using a piece of chalk and a string, the line was bisected. The perpendicular lines were used to lay out a square. A triangular element was laid out on the floor with the perpendicular locations marked. Addition of a plumb weight through the perpendicular line provided a level. Division of a circle around the original intersection provided a rough protractor (Figure 6.21).

These tools were used to lay out the model apse of the cathedral. The string and chalk exercise continued to develop the layout of the stones that would be used to create the pointed Gothic arch. The width of the cathedral would be laid out and circular segment drawn to intersect at the apex of the arch. This layout procedure demonstrated how blocks could be rough cut at the quarry. Rough cutting at the quarry reduced handling and shipping costs while fine finishing was completed at the cathedral site. The base of one buttress was fabricated out of foam blocks. The flying buttresses were laid out on foam boards and were representative of six of the major Gothic cathedrals. The project culminated with a "dedication" of the cathedral. The model and descriptive placards remained in place for approximately two months.

6.7.4   ASSESSMENT COMMENTS

The course was successful in generating interest in medieval construction. The coordination between the History Department and Engineering was effective and the two elements of the course meshed

Figure 6.21: Design and fabrication of a working level.

Figure 6.22: Cathedral dedication.

well. Leading the students to "discover" how arches work and how tools could be made accurately with nothing more than a straightedge, string, and chalk was revealing to many. The project attracted a good deal of interest due to its location on the main campus quadrangle and remained up for the entire summer tour and freshman orientation period.

REFERENCES

[1] *ASCE/SEI 7–10 Minimum Design Loads of Buildings and Other Structures*, ASCE, Reston, VA, 2010.

[2] Kenneth S. Deffeyes, *Beyond oil: the view from Hubbert's peak*, New York, Hill and Wang, 2005 198 pg.

CHAPTER 7

Getting Started

The previous design challenges require a substantial time commitment. Motivating the students to think about design problems often requires a small effort to initiate their thinking. The following are physical and thought problems to lead into thinking about design and design issues. The Column Design Challenge can be used for all ages. The maximum load recorded was a freshman engineering student and it exceeded 12,500 pounds using these rules with no limit on the amount of glue. The Column Design Challenge is coordinated with the state science fair and is an opportunity for the students throughout the state to be introduced to the College of Engineering and Applied Science. The satellite and rubber tire problems introduce students to problems requiring very large and very small numbers. They are ideal to instigate discussion on a problem with no conventional solution and to assign follow up discussion or papers to explore the consequences of their findings. It is not unusual to have solutions to these two problems varying by many orders of magnitude.

7.1 H. T. PERSON DESIGN CHALLENGE FOR PRIMARY AND SECONDARY SCHOOLS IN WYOMING

The Challenge: Using a single sheet of copier paper and up to 4 oz. of white (Elmer's) glue, construct a column that has a minimum height of four inches. The column carrying the largest load will be declared the winner.

When: Testing will be conducted during the State Science Fair. [Students can submit entries in person, by their teachers, or by mail with their name, address, and school. Students need not be present to win.]

Testing: Column load tests will be conducted in a structural testing machine or a frame similar to the photo to the right. Students add one brick at a time up to 15 bricks. Any column carrying 10 bricks will be unloaded and retested in a structural testing machine.

7.1.1 ENGINEERING BACKGROUND AND HISTORY

Leonhard Euler (15 April 1707– 18 September 1783) was a pioneering Swiss mathematician and physicist. He made important discoveries in fields as diverse as infinitesimal calculus and graph theory. He also introduced much of the modern mathematical terminology and notation, particularly for mathematical analysis, such as the notion of a mathematical function. He is also renowned for his work in mechanics, fluid dynamics, optics, and astronomy. He is credited with the Euler buckling equation for the strength of columns (adapted from Wikipedia).

7.1.2   WHAT AN EQUATION TELLS YOU ABOUT DESIGN

The Euler Bucking equation, $P_{cr} = \pi^2 EI/L^2$, indicates the load a column can carry before bucking. P_{cr} is the Euler bucking load and is proportional to its modulus of elasticity, E, the placement of the material away from the center of the column, I, and inversely proportional to the length squared, L. The modulus of elasticity is like a spring constant. Since everyone is using copier paper, the modulus is pretty much the same for everyone. So the challenge is to limit the length to just 4 inches and spread the paper out to improve performance. If the paper is spread too far, then it will buckle locally and not be as strong as a more compact design. Hint: Paper is made by a rolling process, so one direction of the paper has a higher modulus of elasticity than the other direction.

7.2   METEOR COLLISION PROBLEM

This is an in-class problem that may be extended to homework with the solution due in the next class period. Students work in three- or four-person teams to develop an answer to the following problem.

> Each year the earth passes through the Perseus meteor shower. At the peak of the shower, meteorites hit the earth's atmosphere at the rate of 100 per hour. The shower lasts approximately four days.

> You are working on a new geostationary satellite design team. The satellite is four feet in diameter and located 22,500 miles above the earth. Your supervisor is concerned that one of these pieces of space debris may damage the $2 billion dollar satellite.

> You are asked to evaluate two questions. First, what is the approximate probability of the satellite being hit by a particle from the Perseus meteor group? Second, is this a problem? For this problem, probability is not a formal calculation but an assessment, like one chance in a million, that there will be a hit.

> You will orally present your answer on the probability of being hit in class. Your presentation will include the approximate probability and your assessment of whether this is a problem. You should include a discussion of your logic, assumptions, and for any additional information needed to complete your assessment.

7.3   TIRE PARTICLE PROBLEM

This is an in-class problem that leads to a short research follow-up activity. Students work in three- or four-person teams to develop an answer. The first part is to be answered in class the day the problem is given. The second part is given in the next class period to allow the students to look up the impacts.

> Firestone recently recalled over six million tires. The obvious ramifications of defective tires are the loss of control of the vehicle followed by the debris generated when the tread flies off.

> In considering these impacts, the Environmental Protection Agency began to wonder about the health hazard of the particulate matter generated by normal tire wear. If you lived close to a freeway would this material cause respiratory or other problems?

> Your team is asked to investigate this problem. The first question is: "What is the size of the particles generated by normal tire wear?"

> As a follow-up question, your team is asked to recommend whether the EPA should issue a warning about tire particulate matter or mandate new criteria for tire design. How does the size compare to other particles? How might these particles compare to known problems such as asbestos?

> Your team will orally present your answer on the size of the particles in class today with a discussion of your logic and your assumptions. For the follow-up question, identify other information that you may need to complete your recommendations. A one-page summary of your findings and a list of your team members will be handed in at the beginning of the next class period.

7.4   WHAT HAPPENED?

7.4.1   WINDMILL COLLAPSE

An interesting discussion problem provides the class the picture below and asks them to work in groups of two to determine what happened. After about five minutes, ask each group to give one reason what caused the collapse and write it on the board. Continue around the class with each group adding one new possibility. When no further ideas come forward, ask each group to compare the total number of possible reasons they had developed with the total number of possibilities on the board. This leads to a discussion of why teamwork is better than individual effort.

7.4.2   BRIDGE ACCIDENT

Using the four photographs, determine the sequence of the failure when the excavator hit the I-70 bridge.

7.4.3   DEVELOPMENT OF STRESS AND STRAIN CURVES

Demonstrations in Mechanics of Materials classes typically test a steel, aluminum, or brass specimen to develop a stress-strain relationship. While instructive, the test requires specialty equipment and

Figure 7.1: Wind Turbine collapse (Photo Courtesy of Jason Shogren, University of Wyoming).

"black boxes" that isolate the student from the mechanics. The following experiment requires some weights and a ruler to accomplish similar results. The experiment is set up in a room with the weights, platforms, and measuring devices available. There is no reason the experiment cannot be conducted by hanging the specimen from the ceiling. The specimen can extend several times its initial length, so a short specimen is preferred.

Objective: In this experiment, you will determine the stress-strain characteristics of an undefined material and calculate the initial modulus of elasticity.

Background: Read the entire memo before beginning.

[All the necessary equipment for this experiment is available on the first floor of the engineering building. The specimen materials, mass units, measuring tapes, and safety glasses are in a parts box on the shelf on the exterior wall. The test can be adapted to any lab that has weights and a tape measure.]

Figure 7.2: I-70 Bridge collision.

You may work individually or in groups of no more than two. If you work in a group, only a single report is required but both names must be on it.

Safety: Wear safety glasses (included in parts box) when conducting this experiment. Keep feet out from under weights.

Prediction:

Pick one of spools to use for your experiment. On the graph below, qualitatively predict what the stress-strain curve for the material will look like.

stress

Specimen Color: _____

Diameter: _____

strain

Experiment:

1. Select a specimen (smaller diameters provide a greater range of response for the mass units provided). Record the color and diameter of the specimen labeled on the spool.

2. Cut off about 2.5 feet of the specimen from the spool. Tie an overhand loop around one end as shown below.

   (a) Grab one end and pull it against itself so that you form a loop. Take the loop you just formed and make an overhand loop back through, as you would if you were tying a knot. Keep the formed loop large enough to fit over the post (Figure 7.3).

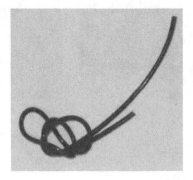

Figure 7.3: Specimen termination.

   (b) Fix this loop securely around the post attached to the base, making sure that the strand is just underneath the washer.

   (c) Tie a small loop in the opposite end the same way. This end drapes over the pulley. Place the mass hanger through the loop you created. A picture of the set-up is shown in Figure 7.4.

3. Place one piece of tape around the material about an inch from the post. Place another piece of tape about five inches from the first. Mark a line on each piece of tape, or on the specimen, to provide a consistent measuring location. Measure and record this initial distance between marks with only the mass hanger in place.

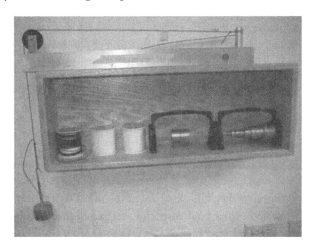

Figure 7.4: Test materials and test setup.

4. Add a mass to the mass hanger and measure the distance between the two pieces of tape. Record both the cumulative mass and distance.

5. Repeat step four, increasing the total mass until the hanger reaches the ground, you run out of weights, or the specimen breaks.

6. Repeat the experiment with a different sequence of mass placement and record the data.

	Mass	Distance		Mass	Distance
1			11		
2			12		
3			13		
4			14		
5			15		
6			16		
7			17		
8			18		
9			19		
10			20		

7. Calculate the stresses for each mass applied to the specimen and the corresponding strain.

(a) Enter your data into Excel and perform the necessary calculations to obtain the stresses and corresponding strains.

(b) Create stress-strain curves for the material using the Excel plot function. Stress should be plotted on the vertical axis and strain on the horizontal axis. Use a scatter plot function with straight lines between data points unless you trust Microsoft to properly interpret the curves between points.

(c) Use the stress-strain plot to determine the initial modulus of elasticity, E.

(Note: $\sigma = \frac{P}{A}, \epsilon = \frac{\Delta L}{L}, E = \frac{\sigma}{\epsilon}$, and mass is not a force)

Furthering the Experiment:

Explain how your prediction compares to the actual stress-strain curve you created. How long did it take to complete the experiment and is this a concern? Does the sequence of loading make a difference? How would you further this experiment to collect more information about the unknown properties of this material? Format your response as a **typed memo (no more than two pages plus the plot)** to your professor and attach your worksheet along with your final Excel sheet and plot, which may be embedded in the memo. This should be a professional quality report.

7.5   NOTES FOR CHAPTER 7

These notes are provided in a separate section so the problem statements may be copied directly. They are not "solutions" but are provided based on the classroom use of the problems.

7.5.1   PAPER COLUMN EXPERIMENT

The paper column experiment typically ends up with loads less than 300 pounds. A simple test frame with about 15 bricks is both visually effective and exciting when the column crushes. We use a small hand-operated universal testing machine for field testing. The hint about the orientation of the paper is to have students think about the consequences of paper alignment but alignment has little effect on the final load carried. This project has worked well in the ES 1000 class as homework with the students bringing the completed columns to class for testing. The explanation of the moment of inertia I is necessarily vague as it is meant to be used by students without a mechanics of materials course. Euler originally defined EI as a combination of material properties and geometry. Young's modulus, E, was not defined until after Euler's death.

7.5.2   METEOR COLLISION PROBLEM

The meteor collision solution requires a change in perspective. Since the meteor showers occur at the same time each year, it is not the meteors that are moving, rather the earth moving through the debris field. One solution to this problem is taking the ratio of the earth's diameter to the diameter of the satellite and multiplying by the number of hits and the duration of the exposure. Because the satellite is in the earth's shadow nearly half the time, a correction can made.

This problem is an education when the student estimates are written on the board followed by each group explaining their methodology. Expect several orders of magnitude difference in the student solutions. Ask which solution is "right." Sometimes the following hints are provided with the problem.

Hints:

What moves, the meteors or the satellite?

What is the density of the meteor particles and what is the volume of the satellite orbit?

7.5.3   TIRE PARTICLE PROBLEM

A solution to this problem requires a larger number of assumptions including the original and final thickness of the tread, the number of miles of tire life and the outside diameter of the tire. From these assumptions an estimated thickness lost in each rotation can be calculated. Assuming each particle is a cube, the size of each piece of dust can be determined. This problem is a good introduction to nano-particles.

The follow-up research requires students to look into the effects of nano-particles. The EPA website has guides for particulate matter size. Students should consider both the size of the particle and the chemical reactivity.

7.5.4   WHAT HAPPENED

Wind Turbine Collapse

The best estimate at the reason for the collapse is that the wind turbine over-sped and one of the blades fractured and hit the tower. No detailed failure mechanism was given. Be prepared for some pretty unusual possible causes. Possible causes ranging from mosquito swarm strikes to UFOs have been suggested.

Bridge Accident

One solution: From the lower left photo, the boom hit the bridge girder just below the parapet wall. Note that the parapet wall is intact. The force of the impact tore the trailer from the tractor, top left, forcing the cab of the equipment to rotate upward until the cab hit the bridge soffit, upper right. The impact bent the boom and extended the hydraulic actuators, top right. When the cab settled, the boom extended well above the deck, lower right.

7.5.5   STRESS-STRAIN CURVES

This experiment is intended to remove all "black boxes" from the determination of a stress-strain curve. The thread in the experiment is a polyurethane string used for kid's snap-bracelets. It is available at Hobby Lobby, Michael's or other craft stores. Larger quantities can be found online. In several years of running this experiment, no student has broken the specimen because the weights hit the floor before the strand ruptures. Even at that, no student report has gone back and shortened the string to determine the breaking capacity.

The snap-bracelet material is interesting because of its extremely high strain to failure. The same experiment can be done with a simple elastic band.

APPENDIX A

H.T. Person Lectures

Year	Speaker	Title	Topic
1998	Daniel P Welsh	Project Manager for Main Street Parade at Walt Disney World	Engineering at Walt Disney World – Making the Magic Happen
1999	Dr. Edward Anderson	Professor, Texas Tech University	Learning in the Digital Age
2000	Dr. Carl Mitchem	Professor, University of Florida	Technology and Ethics: From Expertise to Public Participation
2001	J. Howard Van Boerum	Principal, Van Boerum and Frank Associated	The Design and Construction of the Utah Olympic Park
2002	Lawrence C. Novak	Senior Engineer, Skidmore Owens and Merrill	Perspective from Ground Zero
2003	R. Paul Humberson	Western Area Power Administration	Anatomy of a Blackout
2004	Michael K. Zyskowski	Program manager, Microsoft Flight Simulator project	Microsoft Flight Simulator: The Engineering behind the Game
2005	Joseph A. Anselmi	Mechanical Engineer Aerospace Corporation	GPS: How does it Work?
2006	Neil Kelly	National Renewable Energy Laboratory (NREL) in Golden, Colorado	Engineering Challenges for Future Wind Energy Development
2007	Patrick Tyrell	State Engineer, Cheyenne, Wyoming	Defending our Borders – Protecting Wyoming's Water

2008	Dr. Richard K. Miller	President, Olin College, Needham, Mass.	Education of Engineering Leaders for the 21st Century: Lessons Learned for Re-Invention at Olin College
2009	Lawrence C. Novak	Portland Cement Association	Philosophy of Engineering for the Burj Dubai, the World's Tallest building
2010	Joseph Leimkuhler	Offshore Well Delivery Manager Shell Exploration and Production – Americas	Deepwater Well Design and Operations – Going Forward Post Moratorium
2011	Dr. Daniel Pack	Professor, United States Air Force Academy	Developing Cooperative, Autonomous, and Heterogeneous Unmanned Aerial Vehicles
2012	Governor Mike Sullivan	Former Governor of Wyoming and Ambassador to Ireland	Observations and Reflections on the Benefits and Importance of an Engineering Education

APPENDIX B

Sample Course Syllabus and Policy

ES 1000-1 Introduction to Engineering Study
Course Syllabus and Policy – Fall 2012

Professor: Charles W. Dolan	Peer Assistant: Jane Doe
Office: En 2082	e-mail: xxx@uwyo.edu
e-mail: cdolan@uwyo.edu	
Office Hours: M-W 2:30-4:00 PM	

Catalog Description of Course:

ES1000. Orientation to Engineering Study. 1. [F1<>I, L] Skills and professional development related to engineering. Involves problem solving, critical thinking and ethics, as well as activities to help transition to university environment. Required of all freshmen entering engineering curricula. Students with credit in UNST 1000 may not receive credit for this course. (Normally offered fall semester)

Course Objectives

- ☒ To acquaint students with resources available at the University for their success
- ☒ To get to know some of the faculty and students
- ☒ To help students understand the fields of engineering and the engineering thought process
- ☒ Fulfills USP Requirements for I, L and a portion of the O components
- ☒ To help students to make a successful transition to the University and the College

Week	Topic-WED.	Topic-FRI.	Assignments
1	ES-1000 Information		
1 1	Introduction/Project Professional courtesy and conduct	Introduce classmate using Biographical Sketch; and first thoughts on research topic	**MON** – *Biographical sketch via email to Dolan and Peer Assistant*
Wed.	International Engineering Information session		
2 1	Complete introductions and critique of presentations Information Literacy, Researc Topics	Design Project: Teams and Design Process, "Deep Dive" video	
In class	Provide feedback on oral presentations	Critical Design Issues and schedule	Design teams assigned over weekend
3	Project Brainstorming Shop Safety Video	Library Research Meet in Library 216	*Begin Advisor Interviews*
In class	Critical Issue 1	Critical Issue 2	
4	Design Project: Methodology of Design	Design Day/Shop time	**FRI** – *Research topic due*
	Critical Issue 3		
Sat	Challenge trial run	9 AM-1 PM	Indoor Practice Facility 22nd and Willett Drive

5	Student Success Brainstorming fixes	Academic Policies Professional licensure Advising	
	Critical fix 1	Critical fix 2	
6	Preparing and Giving Oral Presentations	Ethics for Engineers	FRI – *Advisor Interview* *complete TIP tutorial* *completed by midnight* http://tip.uwyo.edu
		Shop time	
7	Research Project Oral Presentations	Research Project Oral Presentations Shop time	MON – *Research Report* *Due by midnight,* *portfolio due in class* *Wednesday* *-Complete ES1000*
Friday	College Open House		Counts as an activity
8	Research Project Oral Presentations	Career Services Knight Hall 222	
Sat	ES-1000 Flight Competition	9 AM - 1 PM	Indoor Practice Facility 22nd and Willett Drive
9	Honors Advising Class evaluations		

INSTRUCTIONAL MATERIALS

Additional Reference material on the web:

`http://wwweng.uwyo.edu/classes/es1000ref/home`

This is the "official" home page of the ES1000 class. It will have much of the information we cover.

The "On-Line Textbook"—*Changes, Challenges & Choices*, Andrea Reeve & Diane LeBlanc (eds.)

`http://www.uwyo.edu/bettergrades/`

on the right hand column. This webpage, though "old," has a lot of great information with just a few broken links.

Library Assignment: `http://tip.uwyo.edu` (Note: there is no `www` in the URL)

Studying Engineering, 3rd edition, Landis, R.B., Discovery Press, L.A., 2007. If you would like a good, all around introduction to the first year and more in engineering, Landis is a good start, cost is about $25.

CLASS REQUIREMENTS
☒ Prepare a Biographical sketch – FRI./Week 1 (Required for O component)
☒ Introduce Research topic – FRI./Week 1 (Required for O component)
☒ Achieve a 70 percent or better on the TIP exam by the end of week 5   ○ Note: Failure to **complete the TIP exam with a 70% or better** will result in an F for a term grade. (Required for L component.)
☒ Complete research topic and assessment papers by week 7   ○ Note: Failure to submit a **research paper,** an **assessment paper,** and a **portfolio** will result in an F for a term grade. (Required for L component.)
☒ **Participate** in a Team in the Design Challenge
☒ **Participate** in Final Oral presentation with group.   Note: Failure to **participate** in Final Oral Presentation will result in an F for a term grade. (Required for O component.)

☒ Complete six outside activities and report by email
   ○ **Required List - four activities**
      1. Advisor interview (by Sept. 18)
      2. Two professional society meetings (one in Sept, one in Oct)
      3. Senior Design Presentations (in December)
   ○ **Elective List - two activities**
      1. One cultural activity (theater, symphony concert, lecture [not a rock concert]), one sports event (football, soccer, swimming, wrestling, etc; must attend the entire event), career fair, departmental presentation, resource fair, or one club activity (in addition to the required meetings, like Habitat, Field trip, etc.) Class meetings and Friday Night Fever don't count.

CLASS POLICY

1. Assignments are due when specified, Late = 0.

2. Class attendance is MANDATORY. University Regulation 6-713 explains how authorized class excuses may be obtained. Missing more than three classes will lower grade by one letter.

GRADING CRITERIA

GRADING CRITERIA		
The course grade will consist of the following points:		
Class attendance and Participation (15 @ 5 pt ea.)	75	
Activities (6 @ 10 pts)	60	(Sr. Des., Advisor, 2 Soc., 2 Elect.)
Biographical sketch	10	
Design Challenge	60	
Library (TIP)	10	
Research Outline	5	
Research Report	40	
Research Source Assessment	20	
Research portfolio	10	
Research oral outline	5	
Research oral presentation	25	
Total Possible Points	320	

Grades							
A	=	90	-	100%	e.g.	>= 288	points
B	=	80	-	89%			
C	=	70	-	79%			
D	=	60	-	60 - 69%			

ASSIGNMENT FORMAT FOR REPORTING ACTIVITIES

1. Send a report of each activity you participated in by email to *both* the PA and to the instructor. The "Send To:" line should read
"YourPA@uwyo.edu; YourProfessor@uwyo.edu"

2. The "Subject:" line *must* start with:
ES1000-XX and then indicate the purpose of the email, i.e., "ES1000-11 - Advisor Interview"

3. The report must be sent **within three days** of the event (i.e., Friday event, send by Monday). No late reports accepted. Only the Senior Design report will be accepted after Oct 19.

4. The report must be *at least* one coherent paragraph, using correct spelling and grammar (one or two lines do not make a coherent paragraph).

5. Content should contain: What you attended, who, when and where the event was held, and what you found to be of interest. Not knowing the name of the speaker or organization is not acceptable.

Examples of two society meeting reports which would also be typical of an elective report:
I attended the American Society of Civil Engineers meeting on Wednesday, February 4. The speaker was James Johnson, a civil engineer from Laramie. The topic was the development of a neighborhood and the various aspects that go into land development. He provided a detailed presentation on the project, addressing such issues as the layout of the neighborhood, plumbing, landscaping, and various legislative aspects of civil engineering.
On Wednesday, February 25th at 5:00 p.m. in engineering building room number 3044, I attended an ITE (Institute of Transportation Engineers) meeting. The speaker was Tammy Reed from Trihedral Corporation which is an environmental and engineering firm located here in Laramie. Ms. Reed talked about a street project for the City of Laramie that will include a new sewer system and reconstruction of the street with medians. The most interesting part about this meeting was that they provided food, which was obviously a tactic to trigger people to come.
Senior Design and advisor reports should be appropriately longer.

- "Disability **Statement:** If you have a physical, learning, or psychological disability and require accommodations, please let the instructor know as soon as possible. You must register with, and provide documentation of your disability to University Disability Support Services (UDSS) in SEO, room 330 Knight Hall." (University Statement) Appropriate protocols will be developed after that time.

- "Academic **Honesty:** The University of Wyoming is built upon a strong foundation of integrity, respect and trust. All members of the university community have a responsibility to be honest and the right to expect honesty from others. Any form of academic dishonesty is unacceptable to our community and will not be tolerated" [from the UW General Bulletin]. Teachers and students should report suspected violations of standards of academic honesty to the instructor, department head, or dean. Other University regulations can be found at: http://www.uwyo.edu/generalcounsel/info.asp?p=3051 (University Statement)

- Academic Dishonesty is any use of *any* work other than your own or of using the same work in two classes. Any infraction of this nature will be pursued to the full extent allowed by University Regulation 6-802 or its successors. This does not disallow working together in groups. It does disallow copying homework within a group which you did not do. Example: Three people cannot do one problem each and share answers. Three people can work together to solve the problems and report them separately. It would be a good idea to read this regulation now.

- Team projects should be worked on jointly. Members of the teams will be required to report on how the team and its members functioned together.

RESEARCH PAPER

Find an application associated with the Design Challenge. Assess the state-of-the-art of that application and why the Challenge is relevant. For example, in robotics examine conditions that include working in hazardous environments, under sea, space, toxic or radioactive sites. You may address this problem from any angle of engineering or computer science including application, design, fabrication, or artificial intelligence.

APPENDIX C

Information Literacy Paper

INFORMATION LITERACY (L) RESEARCH AND ASSESSMENT PAPERS

Embedded in ES 1000

Fall 2012

DEFINITION: Information Literacy is the ability to "recognize when information is needed and to locate, evaluate, and use effectively the needed information." (American Library Association and University Studies Literacy Document)

OBJECTIVE: The objectives of the information literacy component in ES 1000 are several: to learn how to pose a research question, conduct a search on literature that will assist you in answering your question, present a written evaluation of your sources' validity or usefulness, and prepare a written report on your findings.

RESEARCH QUESTION REQUIREMENTS:

You will pass the library TIP (Tutorial) quiz with a 70 or better.

You will select a research question from the list of questions provided in class. A *research question* asks for validation of an idea or why or how something works or behaves in a particular manner. The *objective* is a statement of what is to be examined or demonstrated. For example: a research question may ask, "What is the longest span that a bridge can be constructed?" Answering this requires an understanding of engineering principles, loads, and materials. An objective may be that you will limit your research to suspension bridges as these have been the longest types of construction recorded. You may further refine your objective, such as, "I plan to examine the main cables in a suspension bridge as they are the critical load carrying elements." A question such as "How long is the longest bridge?" is simple fact finding and not a valid research question. By the same token, a report on the construction of the Great Wall of China does not ask a probing question, but rather asks for historical information.

The research question and objective must be submitted to the ES instructor as indicated on the syllabus by the end of Week 2. The Instructor will assist in assuring that the research question and/or objective is not too broad, too complex, or so esoteric that references may be difficult to find.

Each student must prepare a research paper independently.

Your research paper will be based on four and only four sources: one each from a professional journal, a popular literature source, a web-based source, and one additional from any of these sources. You must select three sources that best support your position and one source that refutes it. Everything in your paper must be referenced to these four sources. Pick them carefully.

You will prepare a written paper that answers your research question. The paper will be at least three full pages long (not two and one half) and contain:

A statement of the approved research question and objective.

A discussion of your findings. This is the body of the paper and answers your research question within the limits set by your objectives.

Proper identification and citation of sources and quotations used to support your discussion.

Conclusions regarding the outcome of your research.

A reference list cited in the same format as the primary technical journal used for your paper. This reference list must be in the paper and properly cited in the text of your work.

Academic Integrity: It's your responsibility to be familiar with UW's policies concerning academic dishonesty, both its definitions and its negative consequences. Details can be found under UW Reg 6-802. For more information, go here:
`http://www.uwyo.edu/generalcounsel/_files/docs/uw-reg-6-802.pdf`

RESEARCH ASSESSMENT PAPER

You will prepare a second report at least two full pages long that critically assesses the material you used to select references to prepare your research question report. This critique will contain:

A summary table of the number of sources found, the number read and the number used for your paper (1 or 2). The table will have three categories, journal papers, popular press, and Internet sources. You must identify at least three *references read* in each category.

Literature Search Summary	Journal Articles	Popular Literature	Website
Number of Sources Found			
Number of Sources Read			
Number of Sources Used			

The references for the source material used. Note that the sources selected for the research paper must be repeated in this and the research paper.

An abstract, in your own words, of each source article read for the research report (four). The abstract is approximately one paragraph and restates the most important features of the article for your use.

A critical assessment of the content of each source you read, including a comparison of common features and critical differences. *The assessment must include why the final reference used in your paper is selected.*

PAPER FORMAT

Your papers will be typed, double spaced with 1" margins all around. Type font will be 12 point, Times Roman. Your name and section number will appear on the top right of the first page. The entire project will be submitted in a paper, two pockets, folder, about 9"x12".

Grading: Each paper will receive a grade and comments. The intent of grading on this exercise is to assess your understanding of the research process. The comments provide you with an indication of how the writing meets expectations in college level courses.

An A paper contains all of the required elements, the proper references, correct citation format, a clear response to the question, conclusions, and findings, and a critical assessment of the resources.

A C paper typically shows a lack of focus on the research question and has a rambling response to the question, lacks conclusions, and has inadequate or improper references. The assessment is equally lacking in focus and comparison among articles.

An F Paper is indicative of a student who did not bother to read the instructions, has a poorly formed research question, has not answered the question, and provided no logical references to support the answer to the question. The assessment totally misses the objectives.

RESEARCH PORTFOLIO—OPTIONAL BY SECTION

A research portfolio contains copies of the materials used to develop your papers. Several faculty members require that a research portfolio be included with the papers. Check your section syllabus. The portfolio need not be organized in any specific manner but it should include:

Copies of the four articles cited in your paper. (Copy no more than four pages. The first page should have the title and author of the article. If the journal or book name is not on that sheet, copy a fifth page with the cover of the journal or the copyright page of the book.)

Notes developed during your research sessions.

Specific notes and references to sections in your paper.

DEFINITIONS AND HELP SOURCES

Journals: A journal is a record of transactions maintained by a deliberative body. Contents of a journal are typically peer reviewed and are archived by libraries. That means the articles in the journal are reviewed by two or more people familiar with the subject matter and a judgment is made that the content conforms to established practice. Electronic versions of journals are still considered as journals even if they are found on the Internet. They have the same content and review as the paper version. If the term "Journal" is not in the title, use the library resources to verify it has a journal format.

Popular Literature: Articles in this category are typically authored by a single person and reviewed by an editor for grammar, for libel issues, and for consistency with the editorial objectives. These include newspapers and magazines like *Popular Science*. For this exercise, books are considered popular literature. (In fact, many books undergo a considerable peer review process. The objective is to have you use the search engines available for research work and to examine the content of shorter articles, not to use book as references. If you use a book, you must abstract each chapter that you review.)

Internet Articles: Internet articles may be authored by anyone for any purpose. There is no requirement that they be factual, although many are very good, an equal number are truly bad or wrong. Assessing Internet sources requires some basic understanding of the subject material and often requires an exercise to see if the information on the site can be verified by a second source.

Methods of assessing web based sources are located at

http://www.pbs.org/teacherline/courses/tech340/docs/tech340_bull.pdf

Writing resources: The Writing Center in Coe Library offers assistance in developing written materials for this and all University courses. The services and hours for the semester are found at the Writing Center website

http://www.uwyo.edu/ctl/writing-center/

APPENDIX D

Sample Oral Presentation Evaluation Sheets

You are asked to provide a grade of the students' presentation. The page on the reverse of this sheet has an evaluation form and lists the speakers in the order of their appearance. Check the box to indicate your participation in the review and fill out the sheet using the criteria below and place your grade sheet in the box by the exit. Thank you.

If you want to add your name we would appreciate knowing who participates. If you would like to receive an electronic copy of the student work then add your email address and check the box.

Grading	Each item is graded on a 1-4 scale with 1 being poor, 2 fair, 3 good, and 4 excellent
Evaluation area	**Expectation for a grade of 4**
Organization	The concepts and designs are presented in a logical sequence with each point building on previous work
Clarity	Concepts and designs are presented in terms that are clear, well defined, free of jargon, and easily understood
Verbal	The presenter had good verbal skill including eye contact, voice projection, posture, and poise
Response	The presenter was able to respond to questions in a clear, concise manner.

COMMENTS:

MULTIDISCIPLINARY ALL COLLEGE DESIGN PROJECT – EVALUATION SHEET Fall 2009				
Order of Presentation	Organization	Clarity	Verbal	Response
INTRODUCTION				
Leah				
BEETLE BACKGROUND				
Matt r				
WOOD AVAILABLITY				
Allysa				
Jonathan				
ENERGY PLANT SITE CHALLENGES				
Kolter				
ENERGY PLANT OPTIONS and CO-FIRING				
Jordan				
OPTIONS EVALUATION				
Leah				
WOOD QUALITY ENERGY and ENVIRONMENTAL CONSIDERATIONS				
Jordan				
Matt				
OFFSITE WOOD STORAGE AND HANDLING				
Jon				
ENERGY PLANT WOOD TRANSFER				
Sam				
ENERGY PLANT SITE DEVELOPMENT				
Kolter				
SILO DESIGN				
Allysa				
Dan				
WOOD CHIP HANDLING				
Shane				
COST ANALYSIS AND QUESTIONS				
Leah				

Author's Biography

CHARLES W. DOLAN

Dr. Charles W. Dolan is the first permanent H. T. Person Chair of Engineering at the University of Wyoming. He received his BS in Civil Engineering from the University of Massachusetts and his Masters and Doctorate in Civil Engineering from Cornell University. Dr. Dolan has over 20 years of design experience as a consulting engineer and an additional 25 years of teaching experience. His design projects include the original people mover guideway at the Dallas–Fort Worth airport, the Detroit downtown people mover guideway, the Walt Disney World monorail, and the conceptual design of the Vancouver British Columbia Skytrain and the monorail running down the spine of the Palm Island in Dubai. He has taught at Cornell University, the University of Delaware, and has been involved in teaching classes at Seattle University and the University of Washington prior to joining the faculty at the University of Wyoming.

The H. T. Person Chair is the first endowed chair at the University of Wyoming College of Engineering and Applied Science and focuses on undergraduate education. For over a decade Dr. Dolan has developed the engineering design challenges for the first-year *Introduction to Engineering* course and for a number of years he taught interdisciplinary senior design projects. In addition to conducting interdisciplinary senior design projects, Dr. Dolan is actively engaged in capstone design courses for civil and architectural engineers with a focus on concrete and prestressed concrete structures. He teaches courses on Society and Technology for the University Honors Program. He chaired the UW Read, first-year common reading committee, and served as Department Head and on Tenure and Promotion committees.

Dr. Dolan is co-author of the book *Design of Concrete Structures* with David Darwin and Arthur H. Nilson and serves on the American Concrete Institute Committee 318 *Building Code for Concrete Structures*. He conducts research on the innovative use of prestressed and precast concrete structures and the use of fiber reinforced polymers for strengthening concrete structures. In his research capacity he has served on National Science Foundation committees and edited and contributed to several volumes of work on FRP applications and the durability of FRP strengthening systems and authored numerous technical papers.

Printed in the United States
by Baker & Taylor Publisher Services